DESIGNING THE FUTURE

How Engineering Builds

Creative Critical Thinkers

in the Classroom

ANN KAISER

Solution Tree | Press

Copyright © 2020 by Ann Kaiser

Materials appearing here are copyrighted. With one exception, all rights are reserved. Readers may reproduce only those pages marked "Reproducible." Otherwise, no part of this book may be reproduced or transmitted in any form or by any means (electronic, photocopying, recording, or otherwise) without prior written permission of the publisher.

555 North Morton Street
Bloomington, IN 47404
800.733.6786 (toll free) / 812.336.7700
FAX: 812.336.7790

email: info@SolutionTree.com

SolutionTree.com

Visit **go.SolutionTree.com/21stcenturyskills** to download the free reproducibles in this book.

Printed in the United States of America

Library of Congress Cataloging-in-Publication Data

Names: Kaiser, Ann (Engineering teacher), author.
Title: Designing the future : how engineering builds creative critical thinkers in the classroom / author: Ann Kaiser.
Description: Bloomington, IN : Solution Tree Press, [2019] | Includes bibliographical references and index.
Identifiers: LCCN 2019006823 | ISBN 9781947604551 (perfect bound)
Subjects: LCSH: Engineering design--Study and teaching (Elementary)--Activity programs. | Engineering design--Study and teaching (Secondary)--Activity programs. | Project method in teaching.
Classification: LCC TA147 .K35 2019 | DDC 620/.00420712--dc23
LC record available at https://lccn.loc.gov/2019006823

Solution Tree

Jeffrey C. Jones, CEO
Edmund M. Ackerman, President

Solution Tree Press

President and Publisher: Douglas M. Rife
Associate Publisher: Sarah Payne-Mills
Art Director: Rian Anderson
Managing Production Editor: Kendra Slayton
Senior Production Editor: Tonya Maddox Cupp
Content Development Specialist: Amy Rubenstein
Copy Editor: Evie Madsen
Proofreader: Mark Hain
Text and Cover Designer: Laura Cox
Editorial Assistant: Sarah Ludwig

Acknowledgments

Writing a book like this is a challenge. It is actually a lot like an engineering design project. You think about your intended readers, decide on your goals, and work with the constraints of the space between these covers. You research a lot, consider all sorts of ideas, and then decide what fits those goals and page limits. And then, if you are lucky, you get to collaborate with a team of experts as you engineer your manuscript. I was incredibly fortunate—the wonderful people at Solution Tree are the ultimate book engineers. Everyone I dealt with had vision, creativity, expertise, and empathy. My editors, Amy and Tonya, provided every bit of reflection, feedback, structure, enthusiasm, and hand-holding that I needed. They are masters of modifying to optimize. There were countless times when I wondered what I would have done without them. To everyone at Solution Tree who worked on this project, I can't possibly thank you enough. Your thoughtfulness and professionalism made an incredibly daunting experience manageable and enjoyable. I learned so much from all of you.

This book is for teachers and, in a sense, it was written by the many teachers I have worked with. To all of you who have willingly tried crazy things in workshops, invited me into your classrooms, and shared your experiences—a very heartfelt thank you. And thank you to each of you reading this and to all the teachers I have yet to meet. Your jobs are far more challenging than most people realize and the responsibility of preparing your students for the future is always at the forefront of your commitment. If this book helps you in any small way, I will have accomplished my goal.

The best teacher I know is my colleague, Bryan Colahan. He constantly models what it means to be a reflective, adaptive, and enthusiastic learner. I am grateful for all the support, feedback, and commitment he has shown. Any teacher or student who has the opportunity to learn alongside Bryan has been given a rare gift that will have a lasting impact on them.

I have never heard the phrase "You probably shouldn't try that" from my children or my husband. They are unfailingly supportive and inspiring and I can never thank them enough for that. I think that you reach a point in your life when your heroes live in your future, not your past. Ben, Abby, and Natalie, you are my heroes. As I watch you each adjust fearlessly to the complex world you live in, trying new things, staying true to yourselves, and making a difference, I am often in awe. Thank you for letting me help in any way.

And to Joe, the most creative engineer ever. You are the real inspiration for this book. After watching you design and innovate for years, I walked into my classroom of incredibly bright high school seniors one day about fifteen years ago and said, "Enough." Teaching had to change; my classroom had to be different from the ones we inhabited as students. The young people in that room were the future and they needed to know how to think and embrace new ideas if they were to design the world beyond their classroom. And so, how I taught changed and I have never looked back. Thank you, Joe, for always creating what comes next. Lme.

Solution Tree Press would like to thank the following reviewers:

Kristin Donley
Science Teacher
Monarch High School
Denver, Colorado

Vicky Gorman
Science Teacher
2015 PAEMST Awardee
Medford Memorial Middle School
Medford, New Jersey

Trissa McCabe
Mathematics Teacher
Reno Valley Middle School
Hutchinson, Kansas

Kari Newman
Science Teacher
Durham Academy
Durham, North Carolina

Neil Plotnick
Special Education and
 Computer Science Teacher
Everett High School
Everett, Massachusetts

Phil Stringer
Mathematics Department Head
Crofton House School
Vancouver, British Columbia, Canada

Lauren Zarandona
Mathematics Teacher
The Mississippi School for Mathematics
 and Science
Columbus, Mississippi

Visit **go.SolutionTree.com/21stcenturyskills** to download the free reproducibles in this book.

Table of Contents

Reproducible pages are in italics.

About the Author vii

Introduction 1
    Why Engineering Works for All Subjects 2
    This Book's Organization 3

Part I: Preparation 5

1 Building an Engineering Design Culture 7
    Approaching Change in Three Layers 8
    Thinking Like an Engineer 9
    Forming a Vision by Choosing a Direction 21
    Going Forward 26

2 Deconstructing the Engineering Design Process 27
    Examining the Engineering Design Process 28
    Following the Engineering Design Process Steps 32
    Going Forward 58

3 Designing Projects 59
    Choosing a Challenge 60
    Ensuring Content, Skills, and Process in Every Challenge 61
    Increasing Synergy With Teamwork 70
    Documenting the Process With the Engineering Notebook 78
    Assessing 80
    Going Forward 91

Part II: Activities and Projects 93

4 Starting With Activities That Support Engineering Thinking and Skills ... 95
    Learn From Failure 96
    Know Your Problem 103
    Know Your Options 110
    Develop a Solution 116
    Going Forward 124

5 Introducing Projects for Elementary School 125
- Best Practices for Elementary Students 126
- Overall Approach 127
- English Language Arts–Based Projects 128
- Mathematics-Based Projects 135
- STEAM-Based Projects 141
- Going Forward 147

6 Introducing Projects for Middle and High School 149
- Best Practices for Middle and High School Students 150
- Overall Approach 151
- STEAM Up Front 151
- Really Making It Real-World Global Challenges 162
- Engineering Enablement 171
- Going Forward 180

Part III: Reflection 181

7 Reflecting On, Revising, and Optimizing Your Curriculum 183
- Reflection: The Luxury of Learning 183
- Revision: If It's Broke, Fix It 187
- Optimization: Re-Engineering to Move Toward Sustainable Change in Your Classroom 190
- Going Forward 193

Epilogue 195

Appendix A: Action Plan Summary 199
- *Designing the Future—Action Plan* 200

Appendix B: Project Planning 205
- *Project-Planning Template* 206

Appendix C: Engineering Notebook Forms 213
- *Engineering Notebook Checklist* 214
- *Brainstorming Summary* 215
- *Design Ranking* 216
- *Initial Design Plan* 217
- *Project Task Planner* 219
- *Daily Summary* 220
- *Design Modification Request* 221
- *Final Design Summary* 222

References and Resources 227

Index 235

About the Author

Ann Kaiser is the founder and chief executive officer of ProjectEngin, a STEM education consulting firm dedicated to including engineering design in K–12 classrooms. She is a former engineer and educator with over thirty years of experience. Following her award-winning teaching career, Ann has provided professional development and curriculum support to schools and educators interested in using engineering design thinking and practices to frame active project-based learning (PBL). In addition to workshops, she presents at numerous U.S. and international conferences.

Ann's engineering experience in product and market development in the metals industry led her to value creativity, collaboration, and problem-solving skills as much as to her technical knowledge. As an educator, she worked extensively to include more inquiry and PBL in her mathematics and physics classes. In 2010, she designed and implemented a successful engineering design elective that numerous schools have adopted. This elective introduces a wide range of students to the potential of engineering as both a problem-solving process and a potential career path.

In 2013, Ann won a Fulbright Distinguished Award in Teaching and spent six months in Singapore researching the use and impact of engineering design performance tasks and projects in secondary physics. In 2014, she was named a Top Overseas Teacher by the Academy of Singapore Teachers, Ministry of Education. Ann's innovative work with ProjectEngin led to the company being named the 2017 Small Business Administration Microenterprise of the Year for the New England region.

Ann holds a bachelor of science in engineering from Columbia University and is a member of Tau Beta Pi, the national engineering honor society. As an International Fellow at Columbia's School of International and Public Affairs, she earned a master's degree in international affairs, with a focus on technology transfer.

Visit www.projectengin.com or https://projectengin.wordpress.com to learn more about Ann's work, or follow her on Twitter @ProjectEngin.

To book Ann Kaiser for professional development, contact pd@SolutionTree.com.

Introduction

The exponential growth of information leaves little doubt that school is no longer the exclusive, or even primary, method for acquiring knowledge. Educating young people to deal with a rapidly changing world means they need to learn how to manage a range of ideas and challenges we cannot begin to imagine. It is insufficient to just know facts and ideas; young people must decipher what they need to know and how to use that information. They must be agile lifelong learners, adaptable to new situations, challenges, and careers. Key problem-solving and critical-thinking skills and competencies must be part of every student's education. In addition to increasing content and the inclusion of skills-based learning, new technologies (with learning curves for both teachers and students) make their debut in the classroom every year.

How is one teacher supposed to fit in core content, skills-focused instruction, impactful technology use, differentiated instruction, active learning, and a host of other pedagogical initiatives? It is a herculean task if you think of them as separate endeavors. Layer that over the dynamic environment inherent in the presence, needs, and talents of dozens of students and it is a miracle that teachers ever come back to the classroom after the first weeks of school.

If you've been struggling to find a way to move forward in the face of these moving targets, you have come to the right place. Regardless of what subject or grade you teach, strategic use of engineering design practices and thinking can provide the framework you're looking for. By providing a platform that fosters creative problem solving, engineering design is a powerful tool that can help you *manage* rather than *increase* all aspects of a 21st century education.

The *engineering design process* (EDP) takes students through the stages of defining the problem, researching, brainstorming, prototyping, testing, and optimizing in a way that transcends any one discipline. The EDP provides a framework that values critical thinking, creativity, collaboration, and communication as valuable tools for the application of subject knowledge to develop solutions. Think of the EDP as a vehicle where the passengers of content, skills, technology, and individual needs can jump in for a journey that gets them all to the same place at the same time—to your students. This book gives you resources and activities to help you enhance, not increase, what you are already teaching.

> Regardless of what subject or grade you teach, strategic engineering design practices and thinking provide the framework you're looking for.

Why Engineering Works for All Subjects

Because of its interdisciplinary connections, engineering design can frame projects and activities to provide an authentic way to create challenging learning opportunities at all levels in all classes. No matter what you teach, engineering challenges can encourage creativity and innovation in your classroom—essential skills in the framework for 21st century learning (Partnership for 21st Century Skills, 2008).

Engineering does this because it is synonymous with creativity, innovation, and entrepreneurial thinking. The National Research Council (NRC, 2012) defines *engineering* as a "systematic and often iterative approach to designing objects, processes, and systems to meet human needs and wants" (p. 202). Despite its technical connotations, creativity has always been at the heart of engineering.

Engineering is also at the heart of STEM (science, technology, engineering, and mathematics), as figure I.1 shows. In fact, the Next Generation Science Standards (NGSS Lead States, 2013) include engineering design broadly across the standards for all grade levels, making it inclusive for students who feel traditional science courses marginalize them or fail to provide relevance in what they learn (NRC, 2012). Engineers use science and mathematics to create technologies to meet human needs or solve problems—but we limit ourselves and our students if we confine engineering design to science and mathematics classes.

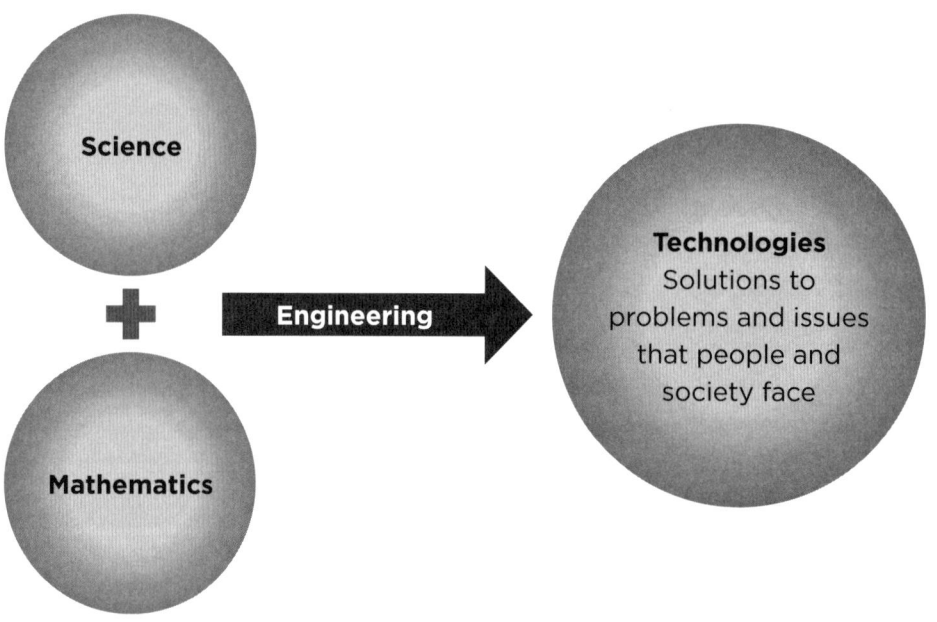

Figure I.1: Making connections in STEM.

When working to meet the needs of people and society, engineers must also have related knowledge of social, historical, economic, and cultural issues. Designing for people often requires understanding visual and aesthetic appeal; we often don't like and won't buy or use what does not look good to us. Good design always requires the skill to communicate, both internally with your team, and externally to inform potential end users. Team members need to communicate with each other to express ideas, explain work, and describe solutions. One of the things I hear most often from students about working on engineering design projects is that they learned how to work with each other. They almost always reference communicating and collaborating. A report by STEMconnector (2018) references data from the Brookings Institution, U.S. Chamber of Commerce, and employer groups indicating collaboration and communication among the skills considered crucial by employers. Because of its many disciplinary connections, engineering design can frame projects and activities to create challenging learning opportunities at all levels in all classes. Most importantly, engineering challenges can help you encourage creativity and innovation in your classroom, no matter what you teach.

> We limit ourselves and our students if we confine engineering design to science and mathematics classes.

Perhaps the most compelling reason to include engineering design throughout K–12 education is the need for students to develop an understanding of the designed world they live in. The Committee on Technological Literacy, National Academy of Engineering (NAE), and NRC (NAE & NRC, 2002) stress the need for increasing *technological literacy*, which they define as "an understanding of the nature and history of technology, a basic hands-on capability related to technology, and an ability to think critically about technological development" (pp. 11–12). We live in a world where countless technologies, from simple pencils to autonomous cars, have been engineered to fulfill our needs, solve our problems, and improve our lives. As educators, we have a responsibility to educate the students who are in our classrooms today in preparation for a highly engineered and complex future.

I have seen engineering design work in hundreds of classrooms. But, it has to work for you and your students. This book means to meet you where *you* are as you begin the process. I provide multiple options—easy-entry, low-risk ideas you can build on to bring out the talent and creativity already in your classroom. This is not a guide for teaching highly technical ideas or creating mini-engineers. And most important, it is not about adding another subject to an already overcrowded educational highway.

This Book's Organization

As you go through this book, choose ideas and activities from each chapter that resonate with your vision and goals. The book moves from culture to curriculum. Part I is where to start if your classroom or school culture needs some work. Chapter 1 details engineering hallmarks and helps you decide a direction for your classroom; chapter 2 explains the EDP. Short (one- to two-class) activities to highlight engineering thinking and parts of the EDP can be found in chapter 4. They are meant to help you shift your classroom culture and are designed to focus on ways of thinking rather than specific curriculum. Most work in a wide range of grade levels and a variety of subject areas. They feature minimal planning and preparation. They are low-risk opportunities to create a little culture shift or to get your feet wet. Longer, multi-lesson projects are the focus of chapters 3, 5, and 6. Projects are centered on a design challenge and require more commitment in terms of time, planning, and resources. Chapter 3 has guidelines and resources for designing your own projects. Chapters 5 and 6 provide numerous descriptions and overviews.

Part I is all about planning and preparing. Part II is about doing, and part III is about making it better. Move between parts I and II as it suits your needs. Part I is a great place to start if you know you need to engineer some changes but are not sure where to begin. If you already have a creative, collaborative learning space and just need ready-made activities and projects, skip to part II. There, chapter 4 gets you started with short activities that support engineering thinking and skills. Chapter 5 introduces projects for elementary school students; chapter 6 introduces projects for middle and high school students.

No matter where you are, part III is important. Chapter 7 helps you reflect on and revise your curriculum for optimization, as well as look toward sustainable change in your classroom. The "Designing the Future—Action Plan" reproducible (pages 200–203) in appendix A can help you craft and organize an approach to move from culture to classroom to projects, and the "Project Task Planner" reproducible (page 219) and other extensive engineering notebook forms are in appendix C.

As you begin using and modifying what works for you, keep the following in mind.

- **The beginning: know what you have and start from where you are.** Change implies you are starting from *something* or *somewhere*. Think about what currently works, what you are comfortable with, and what engages your students. Start there and evolve. Try something small at first, but *do it*!

- **The goal: strive for continuous improvement.** Take time to follow the suggestions in chapter 7 (page 183) and reflect and revise as needed. The best results will come from combining your expertise and skills with the ideas and approaches in this book. Modify, adapt, or discard as needed, but pay careful attention to what does work and how it can be better. *Modify to optimize,* working within your constraints to best meet your criteria or goals.

Most innovation is a team effort. Keep that in mind. I can bring the engineering component into the mix, but *you* are the content expert. You know what you teach and what the learning goals are. The projects I write about in this book are meant to give you initial ideas and you can modify and adapt all of them. Structures as diverse as the human body and massive skyscrapers all obey the same laws of physics. Human needs such as food and shelter are universal; cultures and challenges are often shaped by location and resources. Think about how what you teach is manifested in the designed world beyond your classroom. Use the projects in this book as stepping stones to making those connections.

Try something small at first, but do it!

I have spent half of my career as an engineer and half as an educator. I know why you teach and that is what guides my efforts. This book will help you navigate both worlds. I have used all the ideas, activities, and resources in my workshops, and in my own and my colleagues' classrooms. With the building blocks in this book, you can create a structure that *you* are comfortable with.

PART I
PREPARATION

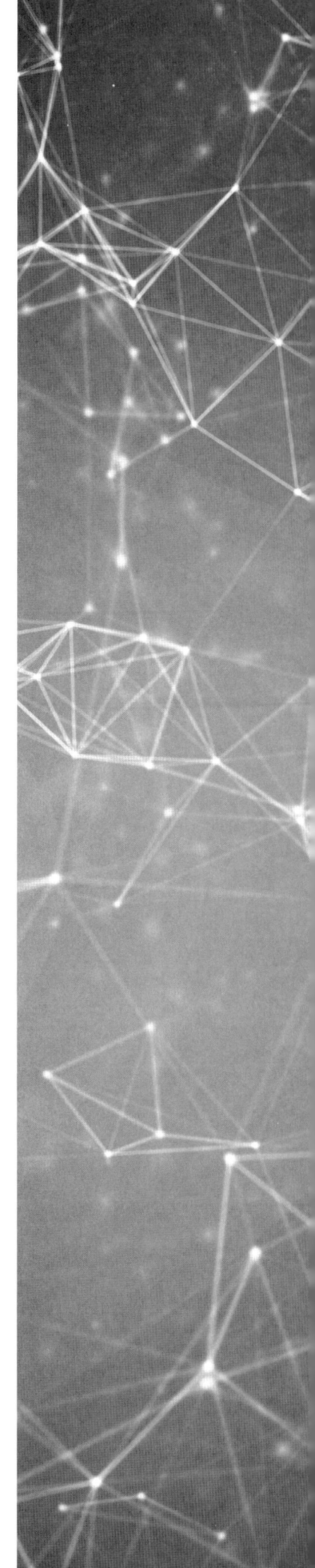

CHAPTER 1

Building an Engineering Design Culture

Without a doubt, the ability to connect the dots is rare, prized and valuable. Connecting dots, solving the problem that hasn't been solved before, seeing the pattern before it is made obvious, is more essential than ever before.

—Seth Godin

Is your classroom culture one of collecting dots or connecting dots? Are your students exploring the messy and complex world around them or moving along a very structured path from A to B to test time? In his book *Creating Innovators*, Tony Wagner (2012) notes:

> Increasingly in the twenty-first century, what you know is far less important than what you can do with what you know. The interest in and ability to create new knowledge to solve new problems is the single most important skill that all students must master today. (p. 142)

Step back and think about how often you, or your students, say the following phrases.

- "It's OK to make mistakes."
- "You probably don't know all you need to know."
- "There's more than one possible answer."
- "Your solution will create some problems."
- "You're not really done."

These phrases are at the heart of the engineering mindset that embraces this type of thinking. If you hope to develop a culture of creative problem solving, these phrases need to become part of the habits of mind in your classroom and school. They go a long way to encouraging dot connecting, not just dot collecting.

This chapter is about how you can develop that vision of a creative, problem-solving culture. The approach to change occurs in three layers from culture to curriculum. You can use short activities to introduce students to engineering design thinking. And, finally, form a vision by choosing the direction that works best for you. Ask yourself how you can help your students move from simply collecting dots, or amassing "more facts, more tests, more need for data," to connecting those dots by applying them to engineering challenges (Godin, 2014).

Approaching Change in Three Layers

Think of change as happening on three layers, with the outermost first.

1. The *where* of your learning space
2. The *how* of your classroom practices
3. The *what* of the curriculum you teach

Figure 1.1 is a visualization of these layers.

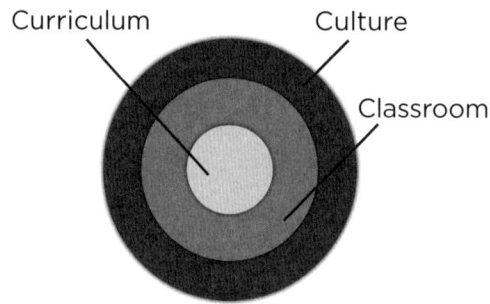

Figure 1.1: Change starts with culture.

Work from the outside in, starting with creating a culture that supports creative problem solving. Begin by following these three steps.

1. Form the foundation for your classroom culture using the hallmarks of an engineering mindset. The hallmarks can help you formulate reasons to include engineering design in your classroom.

2. Move on to classroom activities and practices that highlight different engineering skills and habits of mind to support the classroom culture.

3. Connect large-scale projects to your curriculum. Trying to teach new material in a very new way can overwhelm you and your students, so this step should come last. Try some ideas from chapter 4 (page 95), where the focus is on developing and exploring engineering practices and ways of thinking, before delving into the curriculum-connected project ideas.

At ProjectEngin, we talk about evolution, not revolution. Change is more sustainable when we mimic nature, the ultimate sustainable designer, and take a gradual approach. Think of change as starting with small, shorter *activities* designed to highlight ways of engineering thinking, and then moving on to larger-scale *projects* centered on engineering design challenges. They are more complex, consisting of multiple tasks that require a sense of shared culture and collaboration. Think of the activities suggested in chapter 4 as helping your students develop a helpful toolkit needed to get the best learning experience from a more complex project.

Table 1.1 shows what this culture shift looks like. Where are you now? What changes can encourage the creative problem solvers in your classroom? How can you get where you want to go?

Table 1.1: Shifting the Culture

Cultural Norm	How It Appears in Traditional Education	What It Looks Like in a Creative Problem-Solving Culture
Failure	Not an option; unsuccessful	Learning opportunity
Need to know	Teacher given	Student driven
Solutions	One right answer	Multiple options, optimal design
Thinking paths	Linear, A → B → C	Lateral, systems based
"Are we done?"	"Once you pass the test . . ."	"It can always be better."

Trying to insert an engineering design project into an environment that lacks collaboration, respect, and creativity is rarely successful. It is very much like trying to wear the wrong size shoe. It might work for those first few steps, but you will never get very far without a significant amount of pain.

Based on my observations and feedback from teachers I have worked with, establishing a different classroom culture becomes even more critical as students get older. By fourth or fifth grade, the game of school is coming into focus for most students. They become more comfortable with rote memorization, grades become the benchmark for learning, and they value an efficient linear model of absorption over the exploration and experimentation they were so good at when they were younger. Peer pressure often further reinforces compliance in middle and high school (Burzstyn, Egorov, & Jensen, 2016). There is often little room for mistakes, revision, or collaborative efforts in their mastered world.

Just remember, new approaches layered over traditional practices are rarely sustainable. Be willing to revisit this mindset (and this chapter) when you encounter behaviors and habits that don't support creativity, collaboration, and critical thinking.

Thinking Like an Engineer

Introduce your students to the impact engineering has on their lives with the following activity.

1. Pair students or put them in small groups. (See Increasing Synergy With Teamwork, page 70, for grouping ideas.)
2. Ask the groups to list any ten objects they can see.
3. Next, ask students to list what they think people need to know to make these objects.
4. Share responses in a group discussion, highlighting any obvious connections to engineering, manufacturing, and design. Look for the following.

   - *How pieces are assembled:* You could ask elementary students to identify separate components of one object, and middle or high school students can connect different parts to specific functions or design requirements.

   - *How the objects make it easier to accomplish certain tasks:* This question focuses on the reason for a design. Elementary students can often relate better to what would happen without the device; for instance, how could the teacher write on the board without a marker? You can easily engage middle and high school students in a discussion of how designs evolve to better meet our needs. The cell phone is a classic example of that.

   - *Different styles of the same object:* These items can include pens, chairs, phones, and so on. Observing different designs for the same type of object reinforces the idea that there can be more than one solution to a challenge and that choices often depend on the end user's needs.

It is helpful to introduce students to an engineer's roles before you do any extended topic exploration. Use five- to fifteen-minute videos or TED Talks to hook students. *What Is Engineering?* engages students (National Science Foundation, 2016; https://bit.ly/2O5Wjr6). Use this video to lead a discussion of potential future innovations and some of the issues people tackle with design. Asking students what they think engineers do and what kind of people can be engineers can be a great pre-video discussion. Following up to see if their ideas have changed and discussing some other things engineers do or design would be helpful after viewing the video. In addition, the Royal Academy of Engineering (n.d.; http://thisisengineering.org.uk) videos highlight how engineering plays a role in fields of interest to students, including sports, fashion, music, and global issues.

Be intentional about including some of the following five engineering design hallmarks and start thinking of your classroom as one of dot connectors, not just dot collectors: (1) make mistakes, but don't make the same mistake twice, (2) there is always more to learn, (3) there is no one right answer, (4) your solution will create problems, (5) and you are never done.

Make Mistakes, but Don't Make the Same Mistake Twice

Failure has always been an important part of engineering and innovation. Vicky Hendley (1998), editor for the American Society of Engineering Education's *Prism* publication, says, "Failure can teach students that yesterday's mistakes can lead to today's solutions and tomorrow's innovations."

Failure is an incredibly powerful learning tool. But the learning only happens if you step back, analyze, modify, and then try again. Think of it as *failing forward*, where you learn from failures and use them to your advantage (Maxwell, 2007). Mistakes without reflection and revision waste time and can be self-defeating. We all like the success that comes with moving forward and getting it closer to "right." As a teacher, failure gives you the opportunity to dig into your toolbox and seize all kinds of teachable moments. For example, giving your students a low-risk opportunity to build something like a free-standing tower made out of newspaper (see the Paper Tower of Power activity, page 96) and then testing its ability to support something gives you an engaging way to explain forces, center of mass, and equilibrium as students begin to see patterns in what designs are not so successful. When something works the first time, students are not forced to look beyond the surface. If you have time in a short activity like this to allow for rebuilding, giving students the chance to test and explain their changes, you provide an immediate opportunity for application and concept reinforcement.

> Failure is an incredibly powerful learning tool.

Your own failures and your willingness to learn from them clearly set up a model where you are not the source of all knowledge but more a coach or guide through the messy landscape of things that don't quite work yet.

You may need to shift your own mindset, particularly if you have spent years leading students in science experiments. Be careful not to think of testing and modeling in engineering as quite the same as what has been adopted as the scientific method. There is a significant difference between scientific experimentation and engineering testing. *Science experiments* are designed to provide verification or falsification of a hypothesis. Think of scientific investigation as primarily intended to provide a yes or no answer about a specific hypothesis. *Engineering testing* is designed to assess a solution's functionality, reliability, and overall performance. In this type of testing, the result is often in that gray area of *not quite right*. In other words, failure is often the expected result. The first round of engineering testing informs subsequent modifications. Once you have designed and built your initial prototype, testing it is rarely the end of the process, and testing can involve anything from determining functionality to getting *end user*, or consumer, feedback. Using the EDP, failures lead to design improvements and optimization. If students hope to think like engineers, they need to be comfortable with failure. It is how engineers learn. They will need your help developing the skills and confidence to learn from failure.

A failed approach or solution should not be the basis for grading. Prototype failure should not negatively impact assessment; the ability to learn from failure is what assessment should be based on. If a student can explain what went wrong and connect it to background knowledge, discuss what would improve a design, or rationalize a modification, the student has learned a skill that is far more valuable than creating a successful product the first time around. If you assess a design project on a prototype's aesthetics and successful performance, you risk encouraging a tinkering approach. Students will simply keep trying until something works, with little planning and no real engineering. You can read more in the Assessing section (page 80–91) in chapter 3.

Analyzing failures has always been a valuable learning tool for engineers. Failure analysis is one way engineers improve designs; each prototype is based on what they learned from earlier mistakes. Although it has the advantage of hindsight, failure analysis almost always indicates places where identifying and understanding interactions, connections, and impacts would have been valuable.

This is referred to as *systems thinking*—realizing that the whole is composed of individual parts that are connected, interrelated, and interdependent. You can learn more about systems thinking later in this chapter (pages 18 and 19).

Move your classroom culture from a *no-fail* zone to a *fail forward* (Maxwell, 2007) haven. The best way to establish a culture that supports failure is to use short, low-risk activities as stand-alone investigations or as hands-on hooks for longer-term projects—*quick builds* or *engineering engagement experiences*. They are all limited in time and materials to intentionally create conditions for failure. No assessment, no grades, no risk; just a lot of fun, collaboration, and teachable moments. Some of these activities appear in the Learn From Failure section (page 96) in chapter 4.

Try these activities as part of a curricular unit or as team-building exercises, but always keep the following three things in mind.

1. Limiting materials lets you introduce the idea of constraints and generally encourages more creativity while making failure more likely.

2. Allow no more than twenty minutes depending on the activity. Hands off when you call *time*.

3. Test and debrief each group's attempt with the entire class watching. That way all students witness and learn from several types of failure, and no one group will feel isolated in terms of failing.

Joan Luciano, middle school mathematics teacher at St. John's Academy in Hillsdale, New Jersey, says this about failure:

> Trying something new in the classroom is scary. We have no experience, have limited resources, haven't thought of all the pitfalls or all the questions our students may ask. When we worked with Ann, she would refer to our previous knowledge by pointing out that we are already do so many of these things. We just have to present it from an engineering point of view and stop giving set parameters and outcomes. Encourage failed attempts, since failing is a way to modify, thus getting closer to a solution. These are all life skills as well! (personal communication, November 1, 2018)

There Is Always More to Learn

Think about how you structure a typical curriculum unit. There are several essential questions or big ideas. There may be some formative (or even summative) milestones along the way, nested among the direct instruction, worksheets, and activities you choose. The final destination is generally a summative test, after which it is time to move on to the next topic. While many classroom teachers are modifying this model to include more active learning, differentiated support, and performance-based assessment, it still remains the most common framework.

Unfortunately, many times a summative test signals the end of learning about a certain topic. That is not the case when students are developing a solution to an engineering challenge. If you allow student groups to develop unique solutions, they will almost always need to research and learn more beyond the basic knowledge you provide. This creates a highly authentic model for student-led and lifelong learning.

Don't try to teach every detail that relates to a project. Your job is to give enough direct instruction to enable students to navigate the specifics. For instance, if an engineering project is to design a prosthetic hand that can accomplish one or two tasks for a specific end user, you would need to provide some basic instruction on the hand's overall anatomy, some basic mechanics of levers (how the fingers work), and perhaps some information on prosthetic use. If different teams develop or are assigned specific clients or tasks, they will need to investigate the mechanisms of different hand functions in more depth. For example, if they are trying to help someone throw a ball, they will need to learn different anatomical knowledge than a team whose hand will hold a fork or pencil. In addition, they will need to research how different prostheses manage these different functions.

Think of your role as being a tour guide on a bus, pointing out the key sights and mileposts on the way from challenge to solution. Once you set off, different groups will make different stops and explore different side routes depending on their interest. Some may wander too far into the details, some may come to dead ends, and some may go around in circles. You will often need to bring them back to basic principles and the core route or framework, but it is critical to give them time to get off the bus and explore. Ask students to provide connections and justifications so you can determine whether they are on track, but don't expect every group to follow the same path.

> Think of your role as being a tour guide on a bus, pointing out the key sights and mileposts.

The following approaches will help you identify what students need to know and assess the validity and value of the information they find. Those approaches include providing initial direct instruction that connects to the challenge, using formative assessments of basic concepts for student reflection, always asking for justifications and connections, curating online resources, and including reflection to limit learning by pure trial and error.

Provide Some Initial Direct Instruction and Connect It to the Design Challenge

Teachers who are new to project- or inquiry-based learning often struggle with how much instruction and background material to provide. The most important thing to remember at this point is that it isn't possible to teach everything students will need to complete a project; your goal is to provide a basic framework so they can learn, often under their own power. This does not mean that you should turn them loose to discover everything. Some direct instruction provides a basic mental model or framework for students to build on. When you introduce the initial direct instruction, make sure students see the connections to aspects of the challenge as much as possible. If they see information as potentially useful, they are more likely to be interested in learning about it. You do not need to provide information on every last detail, but you do need to map and fill in the big ideas.

Formatively Assess Basic Concepts to Determine if Background Knowledge Is Sufficient

Project results will be better if you spend some time providing and evaluating background knowledge. Use some preassessments and formative assessments to determine if students have developed a reasonable model of the background curricular content that you are connecting to the project. Give students or groups time to identify what they understand and what they still may need to learn.

I always suggest doing this before students begin their design planning in earnest. Once students have some background knowledge, encourage them to think about what else they may need to know. Determine whether you will need to provide additional direct instruction, if it is possible for students to learn more online, or if it is something that will be clearer as they work with the idea throughout the project.

Requesting that students figure it out works sometimes, but learners can be much more efficient at mastering new concepts if they have a basic model. You're giving them a basic vocabulary before they set out to visit a foreign country. They don't need all the details of grammar; they will add that structure as necessary during their journey. In other words, do not get bogged down in details. Most students will figure out details if they impact their projects, and they will become more fluent in the core concepts you provide as they apply them.

As noted, some teachers find it useful to identify a key point early in the project—generally before students make prototypes—to assess each student's understanding of core concepts. The assessment can be a short quiz, but avoid unnecessary details and allow for retakes. It is important that students show basic understanding of what they need to know to go further. See Assessing (pages 80–91) in chapter 3 for more information.

Always Ask for Justifications and Connections

A great way to evaluate student understanding of key curricular concepts is to consistently ask students for justifications and connections to support key design decisions. That will remind them that this project is not just about making something, but about engineering a solution based on what they know. You can ask for justifications for design decisions and modifications during verbal discussions (making sure to give students plenty of wait time so they can formulate or work through an answer), or have groups document multiple justifications and connections on a graphic organizer or worksheet. The "Initial Design Plan" (pages 217 and 218) and the "Design Modification Request" (page 221) reproducibles reinforce this concept.

The following open-ended questions can get students thinking (Briggs, 2014).

- How does this design decision tie in with what we are learning?
- Why do you think that?
- How do you know?
- Is there a different design choice you might make or have considered? ?
- If you had no limitations, what do you think the best choice would be?

Highlight the need for planning and connections based on key concepts and knowledge about the end user. Asking for those justifications and connections throughout the project keeps the focus on the process, not simply the end product. Losing sight of the design process by simply making something eliminates countless opportunities for building better mental models and learning valuable skills. As you will see throughout this book, it is the EDP, not the finished prototype, that matters.

Asking students to think about some of these types of questions and articulate their answers also encourages them to ask these questions of themselves as they work through solutions. For example, if elementary students are designing camouflage for a photographer and they are supposed to be inspired by nature in a biomimicry project, asking them why they chose certain patterns and where and why that pattern occurs in nature reinforces learning about adaptations. You can ask middle and high school students designing a car that uses a spring-type mousetrap as the "engine" to discuss why they chose larger wheels over smaller ones. They should be able to explain this in terms of rotational inertia, momentum, or distance traveled per rotation, not simply because it "looks good."

Curate Online Resources

Curate some online resources and websites for background information about the challenge and any curricular-related concepts. It is generally more productive to have students use sites you have vetted since the amount and quality of sources can be highly variable. Depending on their age and your policy on online research, students may still need to find some of their own resources, but a curated collection gets them off to a good start. The Research to Learn More (page 41) section includes some basic resources that can work for your students (keeping in mind that online content changes often). Visit **go.SolutionTree.com/21stcenturyskills** for live links to the websites mentioned in this book.

Consider students, for example, who are researching solutions to housing issues in areas that experience monsoonal rains and flooding. You (and students) might want to think about resources that deal with specific tropical and subtropical countries, along with some apps or websites with background information on fluid mechanics and buoyancy. (Chapter 6, page 149, includes a sample overall plan with resources.) Including resources about culture and housing styles might be helpful as well.

Have a discussion with your students about what they think is relevant to their design challenge; address any concerns or direct them to appropriate resources. Post relevant resources in a Google document or on your class website. Provide some support for middle or high school students who may want or need to do some research of their own. In many schools, librarians or tech coordinators are happy to provide an age-appropriate tutorial about searching for and identifying reliable websites. Brita Hammer (2019) has a great set of tips (https://bit.ly/2XAsN3U) for helping students in all grades.

Limit Learning by Pure Trial and Error

Trial and error happens—we all use it at times. Sometimes we need to try different things to see *if* and *how* they work—but you can minimize its impact and turn the work into a more robust learning experience by stressing the need for some justification or prediction *before* making design changes. Simply trying one thing after another does very little to reinforce the application and extension of prior knowledge. If *after* the fact you see that random trying has happened, which is generally the case, encourage students to explain and reflect about why (or why not) something worked then. Watch out for groups that seem to make rapid changes; they are most likely resorting to trial and error.

Asking questions such as the following can help students realize that planning and predicting outcomes *before* a trial might be more helpful and it also allows you to help them make connections after the fact, turning the trial-and-error attempt into a learning experience.

- Why did you choose to try that?
- Did you learn anything before that made you decide to try that?
- Using what you already know, can you explain (or predict) what happened (or will happen)?
- Based on what you know and what you observed happening, what should you do next?

The EDP demands that students tap into the information they know and apply it. Avoid trial and error, but if they absolutely need to try something first, ask for a prediction beforehand and encourage an explanation afterward. And remember that, sometimes, seeing is believing! Your goal should be to identify what students believe after they see what works.

Allow Time for Research

Always allow time for research and remember that some research may involve interviews with and observations of the end user. In all long-term projects, students work in groups or teams. They always have specific roles, or jobs, just like a real design team. Examples of jobs and their responsibilities can be found with the projects in chapters 5 (page 125) and 6 (page 149) and in table 2.1 (page 42). Distribute background research to mimic a team of experts' real-world model. This research generally centers on more detailed information about key concepts you have introduced, things that have been tried before, or an understanding of how a problem impacts people, particularly the intended end user. For instance, in a project about housing that can withstand flooding, one student might research a bit more about structures and materials, another might look into areas that are more prone to flooding and why, and a third might identify housing design and connections to culture in the part of the world that the project focuses on.

Many teachers assign this for homework; they have students compile a set amount of key bulleted points from the website. For grades 1–3, this research may be as simple as a compilation of pictures illustrating typical approaches or patterns. Another example of research for first-grade students is to identify different building materials as well as shapes and uppercase letters in structures around them. In a sense, this creates the student's expertise in his or her given role on the team. Encourage students to divide and conquer research this way throughout the project.

Students can then share their findings with the rest of their group. They can use a simple form to summarize this group research with individual student summaries attached. Have students keep this in the group engineering notebook (page 78). This is all part of the background knowledge needed before designing a solution. I talk about this more in the Know Your Problem (page 103) section in chapter 4.

There Is No One Right Answer

This approach is very different from the traditional model for success in education, where the solution or answer must generally meet a highly constrained and specified description. Often, the answer is binary—it's either right or wrong. Additionally, rather than focusing on solving a problem, students are really on a search for that one answer that leads to a grade of *A*. Convincing your students that real problems rarely have one solution can be a challenge, particularly as they get older and more experienced at finding that one right answer. In some cases, the fact that you do not have an answer key may even prove challenging to you. Although, in an effort to understand student thinking, we might look at why an answer is incorrect, we often start out categorizing it as either right or wrong based on an answer key. This does not work when new solutions are engineered. If you needed a surface to sit on, you probably would not identify one chair as right or wrong. Just as students need to leave the safety of that one right answer behind, you will need to leave the safety of right or wrong when you assess solutions.

Most engineered solutions are *optimized*. In other words, they represent the best options given the *constraints* (or limitations), *criteria* (or goals), and the need to minimize unintended consequences. The simple fact is, if you are creating a solution that involves judging how well it meets several parameters that may be described with a range of values, multiple options can be successful.

This open-endedness creates a much more realistic scenario for students. Point out to them how many options a large shoe store has available to meet the challenge of covering people's feet. Asking students to identify why so many options exist is a great way to have them explore how constraints and criteria lead to the creation of so many solutions. Ask them to consider what happens if you need to buy a new car.

Even with constraints based on cost and safety, you have many possible solutions to your challenge. The optimal solution comes into focus when you add a few more constraints and criteria, particularly those that focus on the end user. When you look at the costs, how you will use the car, your personal preferences for color and style, along with a range of other constraints and criteria, you narrow things down a bit. You could probably come up with several choices but, in the end, you will most likely settle on the one that fits your limitations and meets your personal criteria. Giving five people the challenge of buying a new car will probably result in five different purchases, even if they share the same financial restriction. The range of other factors that influence the decision makes the optimal solution different for each end user.

The following activity conveys the idea of multiple solutions. Ask students to pack a hypothetical backpack for a trip. You specify where they're going and for how long. Give them bag size and overall weight constraints. What they choose to pack will depend on their preferences and plans, illustrating the idea that criteria often connect to end user needs. Each student will likely have different things in his or her backpack, making it clear there is more than one solution to the challenge. An added bonus is that they will learn a bit about scale, estimation, and visualization in the process. Plus, sharing their solutions in class is a great way for students to get to know each other and what matters to each of them. This goes a long way in creating a classroom culture based on collaboration and a willingness to be open to all possible solutions.

This type of activity is also a great way to introduce the idea that we all engineer every day; we are often faced with scenarios that have constraints and, after thinking about the available options, we choose the best solutions. You can convey this message and the idea that real problems rarely have one solution by saying a refrigerator has specific ingredients and asking groups to construct a meal or dish. Simply show a slide of an open refrigerator into which you have pasted a range of ingredients. (Between six and ten ingredients works well.) You are guaranteed to get different menu choices—all focused on the simple challenge of creating something to eat. It also makes it clear that multiple options can be satisfactory—and maybe even tasty!

> We all engineer every day; we are often faced with scenarios that have constraints and, after thinking about the available options, we choose the best solutions.

In addition, you can connect this activity about engineering a meal into a chance to highlight nutrition, different cultures, or even solving a problem that a story character has. It also gives you a way to highlight the fact that even though the constraints were the same (what was in refrigerator), the criteria (what we like to eat) are probably different for everyone and that is what often causes so many different solutions.

Your Solution Will Create Problems

Because of the world's interconnectedness, solutions always create new problems. These are often referred to as *unintended consequences* (Merton, 1936). The hallmark of a good solution is that the positive impacts outweigh the negative consequences. But with more factors involved, outcomes can be increasingly difficult to predict.

For example, in the 1950s, the World Health Organization (2005) began spraying the pesticide DDT in Borneo to get rid of mosquitoes:

> Not long after, the palm-thatched roofs of the village houses began to collapse: a moth larva which fed on the palm fronds had increased because a predatory fly, which ordinarily kept the larva at low levels, had been annihilated by the DDT. The contaminated flies were eaten by lizards which were eaten by housecats which also died. As a result, rats began to invade the dwellings, posing a serious threat of diseases. (p. 6)

While you can find different versions of this story, two things are clear: (1) DDT and other pesticides create all sorts of problems because (2) ecosystems are highly interconnected. A system's existence means actions are rarely highly contained. A solution for malaria led, through many connections, to a potential plague. You can share a short Sustainability Illustrated (2014) video (https://bit.ly/2TOE1zW) about it with students.

Real-world failures allow us to learn from our mistakes. Analyzing them after they have occurred often benefits from hindsight—we can see impacts and unintended consequences once they have happened. Failure prevention is often an exercise in systems thinking, focusing on understanding the connections, interactions, and impacts that result from a design solution before it is put in place. There is a lot that can be learned by examining the causes for historical failures.

Middle and high school students can examine and analyze real-life failure using case studies and online resources about a number of well-known failures, such as the *Challenger* spacecraft explosion (**go.SolutionTree.com/21centuryskills**), the Tacoma Narrows Bridge collapse, and the Hyatt Regency walkway failure. You can find resources on the Engineering Failures website (http://engineeringfailures.org) and also at Engineering.com (https://engineering.com). Failure analysis activities are good ways to introduce both the historical and technical aspects of engineering to students. Visit **go.SolutionTree.com/21stcenturyskills** for case studies about the *Challenger* and the *S.S. Eastland* disasters to share with students.

Most of our interactions with the world are multifaceted, so a change in any one factor can impact a wide range of component and connected parts. Richard K. Miller (2015), president of the Olin College of Engineering, notes many past technologies rose from a narrow focus that underestimated human behavior and interactions. Those technologies have led to many challenges. Electrification changed the world, and in a positive way, but burning fossil fuels to power generators that make the electrical grid a reality has contributed to atmospheric issues and greenhouse gas accumulation. Our ability to move goods over long distances has given us access to many things that make our lives better, but we are only gradually becoming aware of the impact that has on energy and water resources.

A well-designed solution looks at all possible *inputs* (components or actions that go into a product or process), *outputs* (everything that results from a process or product), impacts, and consequences, but it is rare to thoroughly understand every possibility. It is even more challenging in a world of exponentially increasing technology and connections. The process of thinking *holistically* (considering all aspects of a cycle or process) and *laterally* (looking at connections and impacts outside the specified cycle or process) is referred to as *systems thinking*.

You can help students understand this with simple examples such as what they might spend a certain amount of money on. The money goes into the system and what they buy comes out. What is the impact of that purchase in terms of its use, what it means they can't buy, or what it means to the producer or others that they bought it? Students in grades 5–12 are generally highly engaged by the idea of a water footprint for products. (See chapter 6, page 149, for more about water footprint.) It is a great platform for getting students to think about all the ways water is part of a production cycle and how that impacts water availability for other people and products.

Systems thinking is increasingly being identified as a core 21st century skill, and Tracy A. Benson (n.d.), president of the systems-thinking organization Waters Foundation, warns "educators should not underestimate the systems thinking capabilities of children and should re-examine instructional practices that fragment educational objectives into unrelated, non-systemic parts" (p. 2). Encouraging your students to use systems thinking to make design choices and evaluate solutions really does prepare them to be more thoughtful future citizens.

Graphics and visual aids such as *stock-flow maps*, which illustrate how components (stock) move (flow) through a system, and *feedback loop diagrams*, which show how stock changes affect flow, can really bring the idea alive. I like to use the school cafeteria as a way to illustrate stock-flow and feedback loops. Examining how students (stock) move through the lines (flow) can be fun. Looking for changes people make when lines get too long (feedback loops) can also be an engaging way to look at systems thinking. The Waters Foundation (https://waterscenterst.org) has educator and students resources related to systems thinking in the classroom.

You Are Never Done

Always ask your students what might make their designs or ideas better. Encourage them to think about next steps and modifications. It is innovation based on improving and synthesizing existing ideas and products that fuels most product development, not the invention of brand-new, never-seen things. There may be some invention of components along the way, but innovation is at the heart of how people design products and processes for others. Those next steps are what lead to innovation. True innovators know they are never done; they always believe there is a better way.

Two of the biggest technological breakthroughs in the last 150 years—the incandescent lightbulb and the airplane—make it apparent that one is never done.

Thomas Edison lived by the motto "There's a way to do it better—find it!" (as cited in Daum, 2016). Edison filed his first patent for the incandescent lightbulb he built on the technology many others had been working on for over fifty years. His evacuated bulb with a platinum filament burned for a few hours, making it a successful safe alternative to gas lamps. Recognizing the need for longer life, he experimented endlessly with other potential filament materials and filed a second patent one year later for a bulb with a carbonized cotton filament that burned for fifteen hours. He was still not satisfied, and began using a carbonized bamboo filament that lasted for 1,200 hours. This everyday technology has evolved ever since—from tungsten filament incandescent bulbs to fluorescent lighting to modern LED light bulbs with over five times the energy efficiency. All these improvements have come as a result of Edison's search for a better way to get light from materials in a safe, efficient manner (Franklin Institute, n.d.).

Orville and Wilbur Wright built on the work of others and truly followed a well-thought-out design process in the years leading up to the first powered flight in 1903. They continually improved on the glider designs they developed from 1900 to 1902 by researching the work of others, conducting their own wind tunnel tests, and investigating every variable. Even though their first glider flew in 1900, they knew a number of issues would limit future success and so they focused on them methodically. They dealt with lift and then control along all three axes of motion (Smithsonian National Air and Space Museum, n.d.). The Wrights conducted close to one thousand successful glider flights before they were ready to take things to the next level—powered flight (History.com editors, 2009). The Wrights continued to develop their plane, increasing its power and control systems. By 1905, it could fly for a half hour and execute figure eights—practical flight was born (Smithsonian National Air and Space Museum, n.d.). As Orville once said, "If we worked on the assumption that what is accepted as true really is true, there would be little hope for advance" (as cited in History.com Editors, 2009). The Wright brothers' continuous documentation, testing, and modification provide a wonderful early example of the EDP. The way they followed a process to change the world forever can be a fascinating lens for an engineering-based discussion or project in history class.

Ask students to identify the constraints that the Wrights faced and the criteria they set for a successful flight. Then have them look at what was already known about flight and how the Wrights began pulling it all together. Have students list Orville and Wilbur's inspirations, from previous experimenters as well as their own experience. Perhaps the most fascinating part of the process is the Wrights' controlled approach to successive modifications. The Smithsonian National Air and Space Museum (https://s.si.edu/2IGiJ1l) has a wide range of lesson plans and student resources on its website.

As teachers at St. Mary Academy-Bay View in Rhode Island began including some of the activities in this book, students became more engaged; administrators and teachers at all grade levels made changes and saw evidence of new kinds of learning. Lower school principal Margaret Cummings saw a significant change in just five months:

> The teachers have shifted from direct instruction to a more active form of learning. The process is definitely more learner centered, with the teacher as the facilitator. They are posing questions and problems and challenging our students to share their ideas and knowledge with others. The students are taught how to think, to use their imagination, to think critically and to be creative in the process.
>
> The shift from thinking that failing is unacceptable has occurred. They are gaining the confidence to say that it's OK to fail and that trying again leads to success. To hear a ten-year-old girl say that she is proud of herself, after several failed attempts, of completing her task of producing three working electrical circuits; or observing three- and four-year-olds design a structure to include a gear mechanism and then explain the process to me—those are remarkable experiences as an educator! (personal communication, December 19, 2018)

Forming a Vision by Choosing a Direction

How do you get started? Think *backward design*—start with the end (Wiggins & McTighe, 2005). Know where you want your car to go. Envision a destination and head that way. It is challenging, if not impossible, to create a culture shift without a vision in mind and allowing time to embed practices and ways of thinking. The "Designing the Future—Action Plan" reproducible (pages 200–203) can help you create a map for the change you hope to bring to your classroom.

In all my workshops, I start by asking participants to commit to trying just one thing we discuss or model. Inertia and pressure from students are probably your biggest allies in change. Once you start moving, it takes an effort to stop, and generally your students won't let you. They get hooked on the idea that what they are learning has a purpose in designing the world beyond their classroom. Don't overplan or overcommit. There will be some detours and dead ends, but you won't know what they are until you start moving down the road. Being agile and responsive is what innovation is all about and that is the mindset we all need to model for the future.

> Don't overplan or overcommit.

You are actually engineering your curriculum. You can follow the EDP to define your constraints in terms of time and resources and develop the criteria for a successful learning experience. This book will help you consider multiple options. As you try the activities and projects that you think will work, you are prototyping your curriculum and pedagogy. Finally, as you reflect and revise, you are modifying to improve your design. It all starts with a vision of where you want to go.

You may find it useful to choose one or two of the following ideas to help you form a vision as you engineer this new culture: real-world connections, transdisciplinary thinking, skills-based learning, project-based learning (PBL), and student choice and differentiation. Do not try to incorporate too many; in any design scenario, with increasing constraints and criteria, options become limited. Most teachers end up settling on one or two of the following ideas as the primary focus. You will need that vision when you choose what to try, how much time, how many resources you make available, and what your overall plan might look like.

Real-World Connections

Engineering design brings the real world into your classroom. It engages young people in authentic issues and choices, while empowering them to focus on solutions. Engagement increases when students see how an idea or subject is relevant to their lives (Albrecht & Karabenick, 2017; Frymier & Schulman, 1995; Martin & Dowson, 2009). Look around your real world at this moment and list ten things you see. Chances are, many of the objects are part of the built (or designed) world. Now run through a typical day in your classroom. How often do you discuss how the objects and processes so much part of our daily lives were made or evolved?

We all have a highly engineered and complex future. The American Association for the Advancement of Science (AAAS, 1990) notes:

> We have brought the earth to a point where our future well-being will depend heavily on how we develop and use and restrict technology. In turn, that will depend heavily on how well we understand the workings of technology and the social, cultural, economic, and ecological systems within which we live. (p. 107)

Clearly, we cannot hope that our students will be able to tackle the challenges of real-world problems if they lack an understanding of how what they are learning in the classroom connects to the technologies we live with every day. Challenging students to engineer something as simple as a mousetrap car connects some of the physics they learn in school to the basics of automobile design, whether it is gasoline, electrically, or solar powered. Having them follow the EDP to do that in a creative, collaborative way fosters skills that help students through all academics and into careers and personal adult lives.

Transdisciplinary Thinking

Engineering design transcends traditional disciplinary boundaries, creating the foundation for a truly transdisciplinary approach. Real problems do not appear in neat packages confined to one subject area. We live most of our lives beyond school in a trans- and multidisciplinary manner, where getting things done efficiently and well blurs divisions and boundaries. *How* we work is often more impactful than *what* we know. Skills that transcend specific disciplines may be more significant than products and content.

The projects in this book cover a range of disciplines and topics. In addition, the planning materials in chapter 3 (page 59) are designed to work for any subject. I have seen middle school Spanish teachers include EDP in their curricula and have worked with a high school social studies teacher to create a full-year engineering design elective based on global issues.

The EDP approach applies to and often uses ideas from a variety of subjects. A packaging design team engineering a solution to the challenge of minimizing plastic waste might use science to create a more biodegradable material, mathematics to minimize material use and waste, writing skills for labeling and instructions, and art for overall graphics and branding. Engineers use critical thinking (to define problems and evaluate solutions), creativity (to consider new and multiple options), and collaboration (to leverage and synthesize the talents and knowledge of experts from a range of backgrounds). These skills should be a part of learning in all classrooms.

Perhaps one of the most important reasons to bring the real world into your classroom is to model the holistic, transdisciplinary nature of most problems and challenges. Instead of siloed subjects (mathematics for one hour, English for one hour, biology for one hour, and so on) and siloed educators (a teacher may not know the student who struggles with Newton's Laws of Motion is a brilliant writer, or the young person who cannot grasp the meaning of mathematical symbols is a gifted artist), engineering teaches students to adopt problem-solving approaches not specific to any one discipline.

Most importantly, engineering is about solving problems for and improving the lives of people. And people are not defined by classroom subjects. The engineering mindset that there is no one perfect solution and that solutions will create problems (because of interconnectedness) requires an understanding of culture, lifestyles, and perspectives. University of Otago associate professor of higher education Clinton Golding (2009) puts it this way:

> There are various important but complex problems, phenomena and concepts that resist understanding or resolution when approached from single disciplines. Climate change and world poverty are clear examples, but equally, a full understanding of identity, public health, human rights, or knowledge can only be constructed by applying multiple perspectives and ways of thinking. (p. 2)

This interconnectedness also helps students see things from other perspectives, which helps them with end user knowledge and solution creation. Problems rarely come in nice, neat packages with solutions confined to one subject area. We use a lot of associative thinking where one thought leads to another; we connect experiences from the past with novel situations. Boundaries are blurred as we multitask, both physically and cognitively, and we rarely live in isolated one-hour blocks of time.

If you want to reinforce the transdisciplinary nature of engineering, keep the focus on the process—that is where the skills are found. To make the problem's *what* more transdisciplinary, construct a design challenge starting with the subject you are most comfortable with and then try to connect to at least two others. Give your students a way to flex their artistic muscles by creating a logo and appealing packaging for their prototype. Have them venture into the world of technical writing by developing instructions or procedures. Elementary students can learn a lot about planning,

> To make the problem's *what* more transdisciplinary, construct a design challenge starting with the subject you are most comfortable with and then try to connect to at least two others.

budgeting, and mathematics if you create prices for different materials. Combining subjects in one project is often easiest in primary grades, where teachers wear many hats; it often takes a bit of work in middle and high school, but the internet makes information about all subjects available.

Skills-Based Learning

Engineering design creates a culture of innovation that supports creativity, collaboration, communication, and critical thinking (National Academy of Engineering & National Research Council, 2009). These 21st century skills are commonly included in descriptions of engineering *habits of mind*, first referred to by the American Association for the Advancement of Science (1990) and referenced in the NAE and NRC's (2009) report "Engineering in K–12 Education." Engineers are critical thinkers who identify issues and map the constraints and criteria that frame their design space. They are careful to know and understand the end user's needs. Engineers collaborate in teams to creatively generate and investigate potential options, moving skillfully from divergent thinking to the convergent thought processes needed to create a prototype of a possible solution. They optimize solutions to maximize success with minimal negative effects. And engineers must communicate with team members and the outside world to both develop and deploy solutions and technologies. By using engineering design practices and activities in the classroom, a teacher can provide a strong, well-developed framework for modeling and practicing all key 21st century skills (Lucas, Hanson, & Claxson, 2014).

In the majority of cases where I have had the opportunity to speak with students after they complete an engineering design project, they cite working with a team as one of the things they learned. They also often state they learned to work through a lot of decisions and enjoyed creating something. The skills appear naturally if you follow the EDP.

Veteran high school teacher Joanne Cavera at St. Joseph Regional High School in Montvale, New Jersey, uses engineering design projects in her biology and environmental science classes. The difference she sees now (versus using her earlier, more traditional approach) is that students "are learning to work in teams, communicate more effectively, and be creative by working on engineering projects. They compromise, collaborate, and support each other's ideas" (personal communication, October 11, 2018). In addition, her projects' designs require students to make connections across the curriculum, enriching their understanding of many concepts.

Project-Based Learning

The EDP provides a strong framework for PBL. Both engage students in *minds-on* as well as *hands-on* application of concepts. The EDP facilitates backward design (Wiggins & McTighe, 2005).

Students work from a challenge to the key curricular concepts and other information needed to begin formulating solutions. The EDP aids both teachers and students in valuing process over product since it highlights the skills needed to design an effective solution. Along with employing the strategies and practices the Buck Institute for Education (BIE, 2013; www.pblworks.org) suggests, you can use the EDP as a framework for your project planning and for student project management. Following the EDP helps you to design authentic challenges connected to real-world issues; it allows for student voice and choice as teams work to meet criteria they have developed; and the modification process allows for reflection and revision. You will meet these tenets of so-called *Gold Standard PBL* (BIE, 2013) when you follow the EDP.

There is increasing evidence that PBL is academically impactful (Holm, 2011). High-quality PBL is a messy, student-led experience that can initially challenge both students and teachers. By following, documenting, and iterating the key engineering design steps of problem identification, research, solution generation, prototyping, testing, and modifying, PBL practitioners (students along with teachers) can identify a pathway from challenge to solution. And although the content, context, and goals of specific engineering design projects will be different, the overall EDP framework remains a constant across all projects.

The EDP provides a platform for formative assessment, group cooperation, and authentic work—experiences central to PBL. The EDP is also key to focusing on design as a process and the development of a solution as something far beyond simply making a product. Because of this, the EDP reinforces the idea that PBL is "the main course, not the dessert" (Larmer & Mergendoller, 2011). The EDP can help you ensure the hands-on construction of a product follows a process that keeps minds on; it ensures well-connected *learning* and not simply *making*. If you are trying to move to a PBL approach in your classroom, the EDP can provide a consistent framework as students move from project to project. They will already know the *how* of managing a project and developing a solution, so they will be able to focus more on the *what* of the subject area concepts you are helping them learn.

Student Choice and Differentiation

Engineering design challenges provide a level of student choice that supports entrepreneurship, differentiation, and lifelong learning habits. Engage and empower your learners by giving them some control (Parker, Novak, & Bartell, 2017; Ryan & Deci, 2018).

Remember, one of engineering's hallmarks is that there is never one right answer. Part of the art is finding the optimal solution to meet both your goals and limitations. A high level of choice is inherent in the process of engineering a solution. Just think of how many options you have when deciding to purchase a new car. All the cars will get you from one place to another, but each is designed to meet different constraints and criteria or goals.

Student choice lets you take advantage of the rich tapestry of talents and interests in your classroom. In a world that values an entrepreneurial mindset, any approach that encourages the contributions of unique talents and perspectives can go a long way in developing the skills our students need for future success. We are more creative when ideas go beyond reflecting a central focus or vision. Innovation experts often note that innovation happens at the edges, not in the common central core. Student choice allows for diffused vision and perspective.

Dictating the subject, mode, and outcome of learning—a *compliance* education model—does little to encourage student engagement. A compliance education starts with bells and schedules and ends with standardized assessments. Clearly, these practices cannot support self-direction, intrinsic motivation, and engagement. In fact, student engagement surveys indicate the longer a student is in school, the less enthusiastic they are about learning (Busteed, 2013).

A well-crafted engineering design challenge lays out a basic need to meet or a challenge that needs a solution. Instead of saying saying, "Build a bridge" using specific materials and designing it to hold a certain amount of weight, ask students to "Design a way to cross a hypothetical river safely while efficiently using resources available locally." That creates a scenario where students envision a unique end user, perhaps even a place, and design to meet those needs and criteria. Each group will create a solution that looks different while attempting to meet the same challenge. Although they begin with the same basic conceptual framework, they will also need to research and learn some different concepts and topics in order to understand their end user and planned solution. Different approaches to the same learning challenge create natural, student-led efforts.

Three elements help create opportunities for student choice and differentiation.

1. Craft your challenge well by stating the situation or problem and staying very clear of any one solution. Don't say *bridge* if *crossing a body of water* is the real issue. Don't specify a *wheelchair* if the challenge is really *increased mobility*.

2. Provide a choice of potential end users, or let students choose them. A mobility assistance device for a five-year-old girl might be very different than one for a sixteen-year-old girl.

3. Most importantly, let groups develop their own criteria for a successful design. You might specify one or two, but three or four should be group generated. Since their design should satisfy their criteria, their solution will most likely be different from other groups'. You will know that you are doing this right when all the solutions are different. It's the idea that there are always multiple solutions to a challenge that allow you to provide a learning experience that has student choice and differentiation built in.

Going Forward

Do not increase what you need to do by trying new learning models on top of old approaches. Enhance learning by making some minor adjustments in your culture; more changes will follow. Keeps the words *evolution, innovation,* and *modification* in mind; avoid *revolution, invention,* and *final*. As the engineering culture begins to take root, the things you will want to try will blend into and enhance the learning experience for your students, and for you as well. Students will start to connect more of the dots they collect.

CHAPTER 2

Deconstructing the Engineering Design Process

Scientists discover the world that exists; engineers create the world that never was.

—Theodore von Kármán

How can you channel all that culture-shifting energy to move from challenge to solution? The answer is easy: process, process, process! Use the EDP. Using the EDP to support your classroom culture will make that approach an easy fit as you frame larger-scale projects.

Creativity and problem solving have always been at the heart of engineering design. Most importantly, as Linda Katehi, former chair of the NAE Committee on K–12 Engineering Education, says, "Design is not just an engineering skill. It's a skill for everyone" (as cited in NAE, 2013, p. 21). Engineering design requires innovative thinking, analytical judgment, the power of many minds, and the ability to defend and describe ideas.

Most students enjoy making and building things, but they will often resort to either a trial-and-error approach or a superficial "It looks good." Again, keep the focus on the design process. Doing that enables you to support and assess skill development. In addition, it makes it easier to engage students in the introduction, discussion, and application of curricular content since they have a problem to solve. Following the EDP also ensures students justify their choices, fit modifications to meet constraints and criteria, and use a content-based rationale for design decisions.

This chapter examines the EDP's overall characteristics, connections, and rationale. It helps you follow the EDP steps, providing tips to make it more natural and simpler to apply at all levels. This chapter highlights connections to key 21st century skills and ways of thinking critical for students. Chapter 4 (page 95) continues exploring the EDP with activities you can try in all subjects and classes.

Examining the Engineering Design Process

The EDP is an iterative, open-ended, problem-solving method that lies at the core of all engineering projects. It is the method engineers use to move from initial ideas to a successful product, process, or system (Accreditation Board on Engineering and Technology, 2018). As a creative process that effectively chunks challenges into manageable steps, the EDP starts with problem definition and framing the design challenge in terms of constraints (limitations) and criteria (goals).

The design process then moves through the consideration of multiple options for a solution and the development of an optimized prototype. In identifying this process as the part of engineering most relevant and valuable in K–12 education, the NRC (2012) specifically identifies core ideas as being "how engineering problems are defined and delimited, how models can be used to develop and refine possible solutions to a design problem, and what methods can be employed to optimize a design" (p. 202).

There are many representations of the EDP available from a wide range of sources. They can depict anywhere from three to twelve steps. The one I use most often is in figure 2.1.

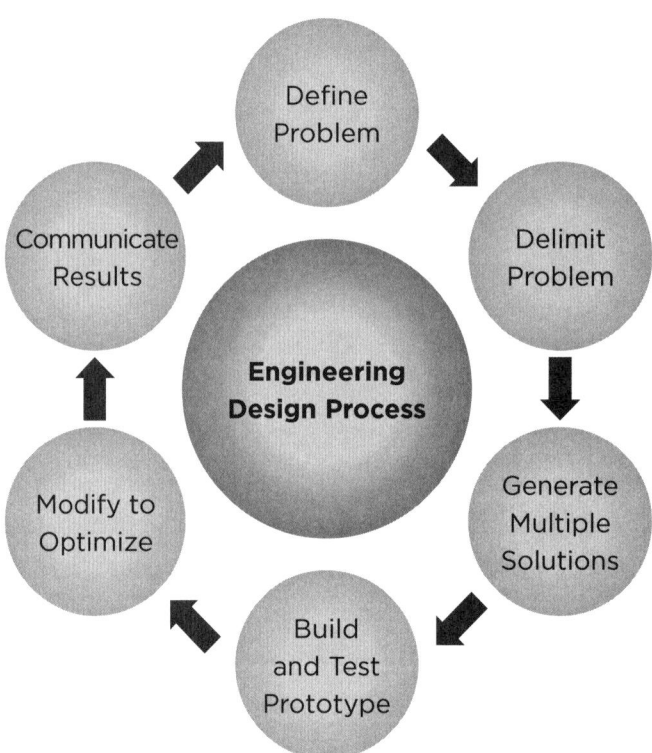

Source: Adapted from TeachEngineering, n.d.

Figure 2.1: Key steps of the EDP.

While this is a process (each step of which I detail in the section titled Following the Engineering Design Process Steps, page 32), I urge you to resist attempts to have students memorize or even follow the steps in a prescriptive manner. Every challenge is different; there are times when some steps are more significant than others or when students need to backtrack. The EDP is almost always iterative, and going back to previous steps often provides valuable insights and modifications. It is not uncommon for students to have to return to the brainstorming phase after testing. Revisiting and analyzing some of the potential solutions they thought about earlier may shed light on potential modifications.

The EDP should function like a framework, not a rigid algorithm. Some of the teachers I work with think of it as a flow, not a process or procedure. Additionally, not every project needs to include everything in depth. In fact, very few projects will include each step in a fully detailed manner. There simply is not enough time in the school year.

> Very few projects will include each step in a fully detailed manner.

Textbook publishers and curriculum designers have somewhat abused and robbed creativity in favor of the idea of one highly scripted scientific method. It is not a "simple recipe for performing scientific investigations. . . . It can easily be misinterpreted as linear and 'cookbook' . . . but science is complex and cannot be reduced to a single, prepackaged recipe" (University of California Museum of Paleontology, n.d., p. 1). Be careful not to let the EDP suffer the same fate in your classroom. Think flow, not script.

I often characterize the EDP as being composed of three phases with different levels of the convergent and divergent thinking needed at each point. Thinking about the EDP in this manner can make the process more intuitive.

Three Phases Within the Process

Using the overall process and concept of engineering design to its best advantage can be challenging if you make black-and-white distinctions between the different steps. Remember, we often call on our inner engineer to solve our problems. It helps, first and foremost, to think of the EDP as a natural problem-solving process first. In thinking about how you can use the EDP to frame a project, think of it as having the following three phases.

1. **Know your problem:** Think of this first phase as the time to generate student engagement and critical thinking.

2. **Know your options:** This is the phase when students can begin to feel empowered and able to create and consider innovative solutions. Creativity is up front in this phase.

3. **Develop a solution:** This phase requires students to complete the process of engineering a suitable solution and sharing their results. Students need to collaborate and communicate throughout this phase.

Most teachers I have worked with find the graphic in figure 2.2 (page 30) helpful as they create activities and projects; it helps them chunk the process into three manageable parts and make connections to concepts. But, always remember that engineering design is an iterative process. Students may have to revisit or repeat any of the steps, and the process can look slightly different for different challenges—or even for different groups tackling the same challenge.

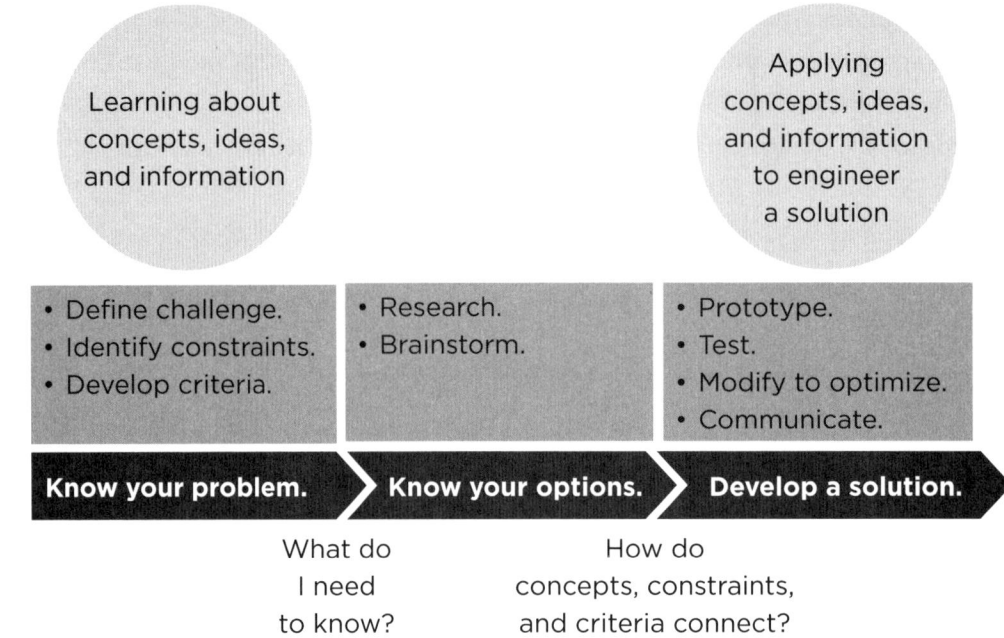

Figure 2.2: Three phases of the EDP.

A group of student athletes designing better athletic shoes may not need as much background research because they can already identify with the end user and are familiar with the product. Students designing solar lighting for regions off the grid will most likely research more. Brainstorming and generating multiple innovative solutions may actually be challenging for our shoe designers since they live in a world almost overloaded by shoe options. The lighting specialists may struggle to overcome constraints if they are designing for people with limited resources in remote regions.

At all grade levels, the Next Generation Science Standards (NGSS) focus on defining problems with constraints and criteria, developing multiple possible solutions, and prototyping and testing to optimize a solution (NGSS Lead States, 2013). As students progress through the grades, they will encounter increasing levels of the following in the NGSS.

- Closer, more explicit examination of constraints and criteria and the factors that impact them
- More robust consideration of solution options, synthesis of ideas, and of characteristics found in various options and solutions
- More rigorous testing and generated data use to better inform modifications and optimized solution development; increased use of systems thinking

A building project in grades K–2 may begin with a basic idea of the limitations (constraints) of classroom materials for building a prototype; criteria may be focused on what looks pleasing. By grade 3, students are more comfortable with more clearly defined criteria and constraints and are more connected to real-world conditions. As students move onto middle school, constraints and criteria are more defined by background knowledge and the end user.

In terms of multiple solutions, elementary students may actually come up with the craziest ideas, but they struggle to find patterns among them and actually settle on one. As students reach middle school, they can organize and analyze the possible solutions they have generated and should be expected to look for common features or ideas that are represented multiple times before moving on to the prototyping phase. They should recognize patterns in various design components, such as common roof designs. They should also synthesize parts of ideas—for instance, combining a materials choice from one idea, the design of another idea, and the size of yet another idea. Synthesis is a significant part of innovation.

When it comes to testing prototypes, the youngest engineers tend to be in a go/no go mindset: Did the building stand up? Could it hold some weight? They should begin to connect some features to success or failure, but mainly in a qualitative manner. Students in grades 3–5 can begin developing more standardized testing procedures and should be able to collect some data.

The next step, as students move through grades 6–8, is to use that data to justify any modifications. High school students can conduct several tests; they might test the structure's ability to support added weight, stay dry in a storm, and resist wind damage. Their analysis should be more complex, depending on the project focus, and they need to look at how any one modification might impact different performance and testing parameters. As students consider these types of impacts, trade-offs, and results, they will naturally move into a more systems thinking approach.

Convergent and Divergent Thinking

Designers may use many iterations as they attempt to develop the optimum solution. The EDP is often characterized as design under constraints to meet specified criteria. Designers consider and evaluate multiple possibilities before choosing a prototype to model or build. As engineers work to optimize the solution, testing the prototype leads to modifications and the consideration of trade-offs and impacts. From an overall skills perspective, the EDP requires students to move back and forth between convergent (arrows coming together) and divergent thinking (arrows splitting apart), thinking both linearly and laterally. See figure 2.3. It is truly exercise for the brain!

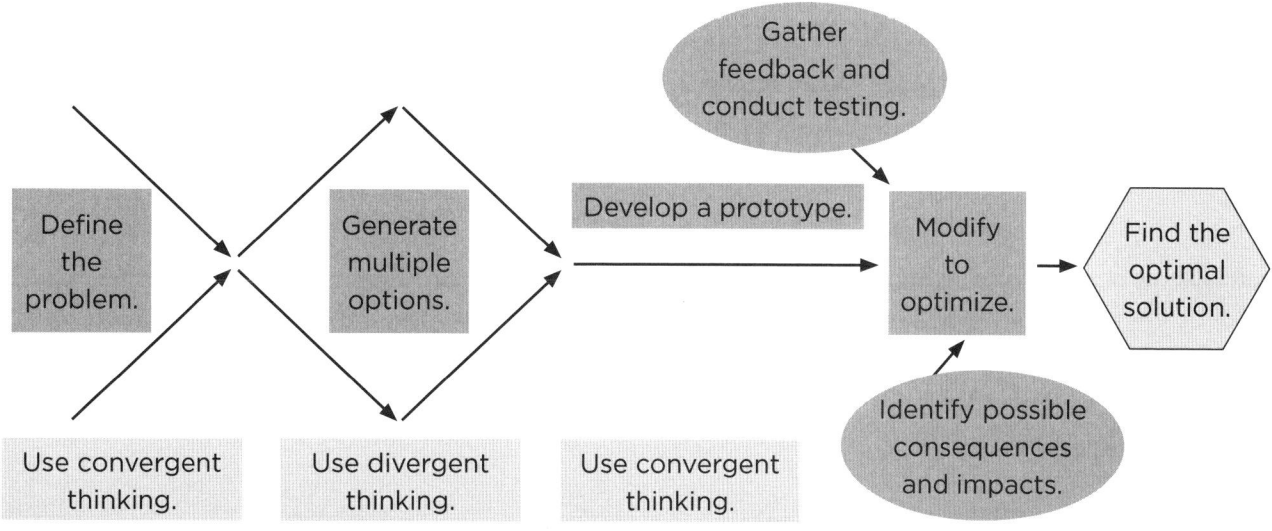

Figure 2.3: Engineering thinking.

Following the Engineering Design Process Steps

The rest of this chapter looks at the EDP steps and three phases: (1) know your problem, (2) know your options (after which you'll take a big-picture look with students), and (3) develop a solution. You'll also find tips and suggestions for making engineering design thinking and practices part of your classroom culture. Chapter 4 (page 95) contains specific activities to help you focus on the skills and thinking involved in each step.

This chapter intentionally does not include specific project content; instead, it allows you to try parts of the EDP as separate activities or use the EDP to support some of the content you already teach. This is where you start building your toolbox of engineering design practices, so you can plug them in as needed in various projects.

To make the following detailed description of the EDP clearer, the various steps will be numbered as follows. The "Designing the Future—Action Plan" reproducible (pages 200–203) helps you visualize and plan around these pieces.

Phase 1: Know your problem.

Step 1. Define your challenge.

Step 2. Understand the end user.

Step 3. Identify constraints and criteria.

Phase 2: Know your options.

Step 4. Research to learn more.

Step 5. Brainstorm possible solutions.

Phase 3: Develop a solution.

Step 6. Decide on a prototype.

Step 7. Create a prototype.

Step 8. Develop and perform tests on prototypes.

Step 9. Modify to optimize.

The final piece is sharing the news.

This numbering is meant to help you work through the process when you are new to it. It is not meant to be prescriptive. You can sometimes develop criteria before researching your end user; you may want to communicate some results for feedback and then find yourself back at step 8, testing; or your understanding of the challenge may be altered a bit after you research to learn more in step 4. What matters more than any step-by-step order is that you and your students spend some time in each of the three phases, thinking critically about the problem, creatively about possible options, and collaboratively about how to reach a solution.

Know Your Problem

During this phase, you and your students will define your challenge, understand the end user, identify constraints and criteria, and take a big-picture look. Think of this first phase as a launch pad. It needs to be well delineated. As a teacher, you cannot help your students undertake any sort of learning journey if they don't know what the starting point is.

As you work through phase one, pay close attention to how students word or reword a challenge. Decide how focused on a specific end user they will be and determine if you will decide who that is, or if students will. Spend time on the idea of constraints and criteria; it is important language and it appears in standards for students as young as grade 3 for a reason. We are more creative when faced with limitations (constraints) and more productive when we have goals (criteria). Work with your students to identify what they need to know in order to deal with this challenge so that they are well-equipped for a learning journey that you hope they will lead themselves on.

Try to use the word *challenge* (or the phrase *design challenge*) more than the word *problem*. In my experience with both teachers and students, *problem* seems to suggest a fairly linear path to *solution*. I suspect that thinking is in part due to the convergent programming in most classrooms and our human need to get people help quickly.

Step One: Define Your Challenge

A Yale University engineering professor is credited as saying "If I had only one hour to solve a problem, I would spend up to two-thirds of that hour in attempting to define what the problem is" (as cited in Markle, 1966). Clearly, you cannot expect to get where you want to go—reach a solution—if you don't know where you are. Defining the problem lets you know where you are. It is important to know the true issue before attempting to develop a solution.

Here is an example: suppose you have an issue with getting to work on time. You need to further define the issue if you hope to work out a viable solution. If the issue is traffic, solutions might involve alternate routes or even a different residence or job. If you hit the snooze button a number of times, you are already behind schedule even if you are driving the sole vehicle on the road, using a flying car, or living next to your workplace. Addressing the wrong issue by developing a different way to commute may not be the best answer. You are more likely the first end user for my new Bed-Eject-O-Matic, and I missed the opportunity!

Always ensure you and your students have a clear understanding of the core problem or challenge. The following will help clarify that understanding.

- Ask questions.
- Look for cause-and-effect relationships.
- Identify impacts.

Engineering design challenges are generally *ill-defined* problems, but they are not *undefined* problems. They are real problems with many possible solutions. In ill-defined problems, numerous factors may be important and it can be challenging to find a path to a solution, since there are many possibilities. It is critical to have a process like the EDP that helps you to rationalize choices and fit solutions to criteria and constraints. Challenges such as access to clean water and providing reliable traffic management are highly multifaceted. They are defined, but there is no one clear path to solution. Even deciding what to have for dinner can be considered an ill-defined problem. You have multiple options—buying some food at the grocery store, eating what you have on hand, getting take-out, or going out to eat. The solution you pick might be different than the one your colleague chooses due to varying constraints of time and money and criteria determined by individual taste, but you both began with the same problem.

Ill-defined problems can be multifaceted and there may not be much known about them at first. The engineer's—the student's—first job is to bring some definition to them by identifying the following (for grades 3 and older).

- What influences and factors matter?
- What is known and what needs to be known?
- What might be important when thinking about a solution?

A graphic organizer or worksheet can be helpful at this point. Focusing on the *five Ws*—who, what, when, where, and why—is a good way to cover all aspects of fine-tuning your problem definition. Figure 2.4 works for middle and high school. Figure 2.5 works for elementary school.

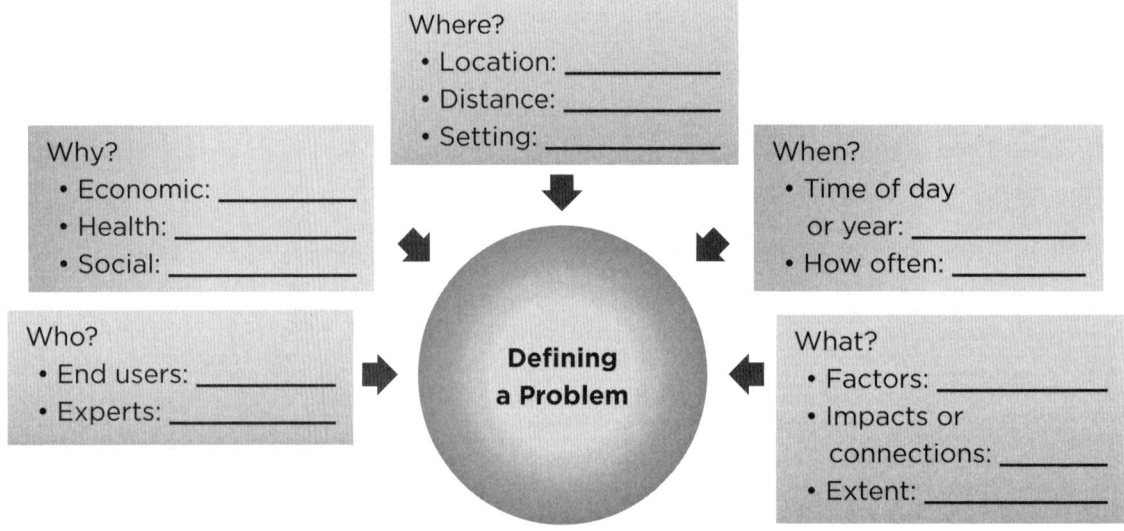

Figure 2.4: The five Ws of problem definition—middle and high school.

Visit **go.SolutionTree.com/21stcenturyskills** *for a free reproducible version of this figure.*

To really understand and define your problem, you need to ask some fundamental questions.

- Why? Think about whether money, health, social, or any other issues have caused the problem. Knowing the cause sets you on the path to solution.
- Where? Different problems occur because of where we live, go to work, or attend school, and other factors.
- When? It is helpful to know how frequent a problem is. A minor problem that happens every day (such as a leaky faucet) can be just as challenging as a major problem that occurs once every few years (such as flooding).

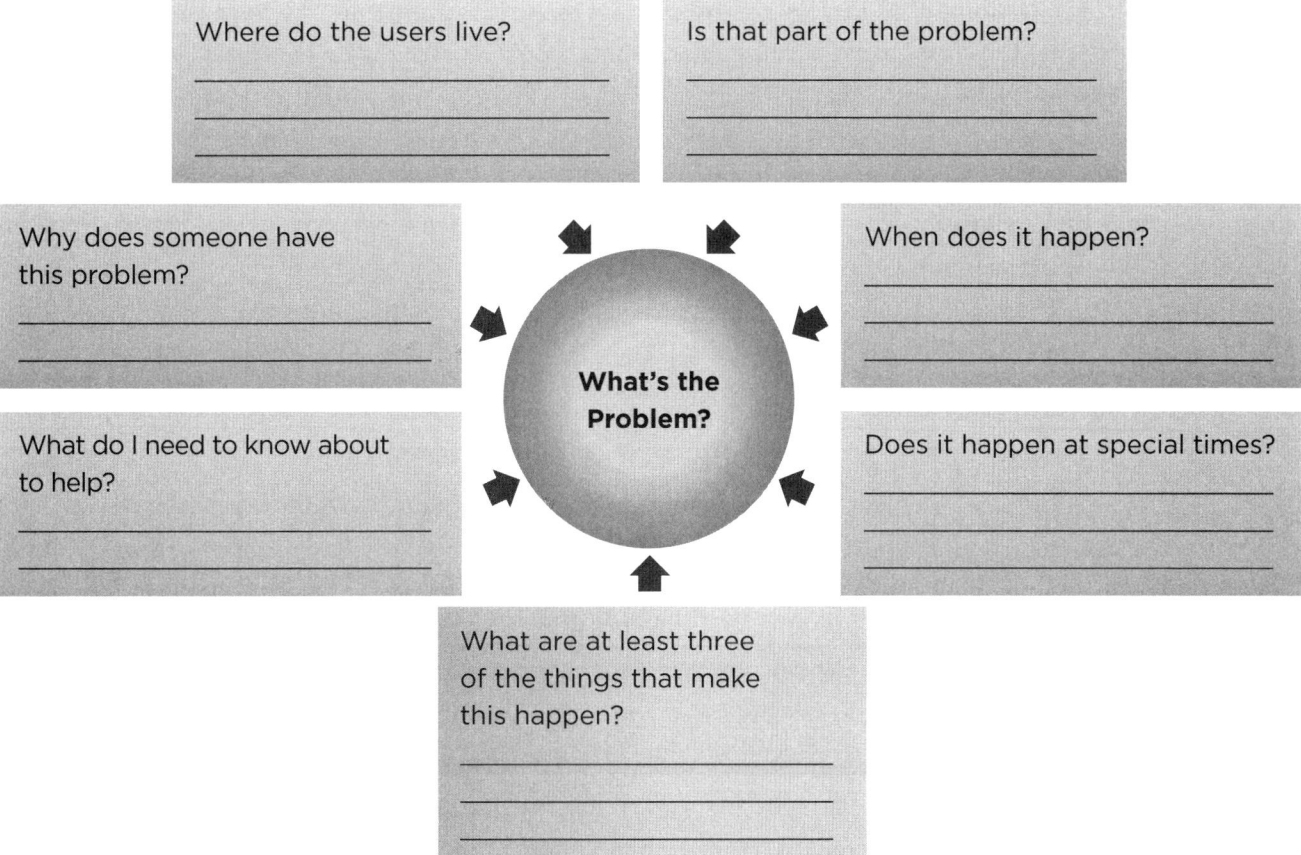

Figure 2.5: The five Ws of problem definition—elementary school.

*Visit **go.SolutionTree.com/21stcenturyskills** for a free reproducible version of this figure.*

- What? Think of all of the ins and outs related to this problem. If you get to work late, who is impacted? What might not bother one person can significantly impact others.
- Who? This is most important. Engineers solve problems for people. Once you have identified who has the problem, it is important to try to understand them more.

Step Two: Understand the End User

The next phase of defining your problem is to know and understand the end user. Developing a new product or solution and *then* searching for an end user is not very effective. The best designs have value to the end user, targeting their needs and addressing their *pain points* (problems).

Let student groups have a good deal of say in terms of who their targeted end user will be by providing some possible choices or allowing them to create a fictional option. This creates the potential for highly empathic design, since students will be committed to helping this person or group of people. This always ensures more ownership of and engagement in the project. Have students think about the age, gender, lifestyle, hobbies, and goals of the potential end user. This part of the background research is almost entirely student driven, but it is helpful to start students with some potential good sources of information (as mentioned in Curate Online Resources, page 15).

Good design hinges on identifying and keeping the end user in mind. Products and processes are engineered to meet someone's needs or to solve people's problems. Knowing who will use the product and what is important to them is a key part of knowing the problem. End users are the focus of human-centered design and design thinking in a wide range of applications, from business to technological development. This EDP step highlights empathy and the idea that observations and interviews are valuable forms of research. Empathy is important for collaboration, and "is a key part of being a responsible and helpful community member at school and elsewhere . . . [and] can also be a route to academic and career success, because it helps people understand and work with others" (Jones, Weissbourd, Bouffard, Kahn, & Ross, 2018).

The Name Your Pain activity (page 105) in chapter 4 fosters the end users' perspective, builds teams, provides valuable interviewing experience, and gives students a chance to vent about what bugs them.

Step Three: Identify Constraints and Criteria

In my experience, many educators and students struggle with managing the EDP because they fail to clearly identify and evaluate constraints and criteria and relate design decisions back to them. Paying attention to and documenting this part of the process provides a framework for a well-designed solution as well as creating a rubric focused on goals and characteristics. You can read more about rubrics in the Assessing section (pages 80–91).

Together, constraints and criteria frame the design space, as in figure 2.6. The two elements bring definition to an ill-defined problem.

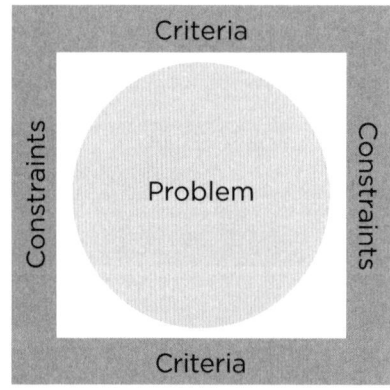

Figure 2.6: Constraints and criteria.

Think of constraints and criteria as a Goldilocks problem. If there are too many, the problem is too well defined to be a true engineering challenge and forces students back to thinking there is one right answer. Think of all those criteria as funneling the group along with very little room for divergent thinking. If there are too few criteria, the space is too big to get started because design becomes an overwhelming moving target. The group's sole focus then shifts to constraints; that can create solutions that look much the same from group to group.

With practice you can find the sweet spot. Finding the right number of constraints and criteria takes experience and patience. Figure 2.7 shows that challenge graphically.

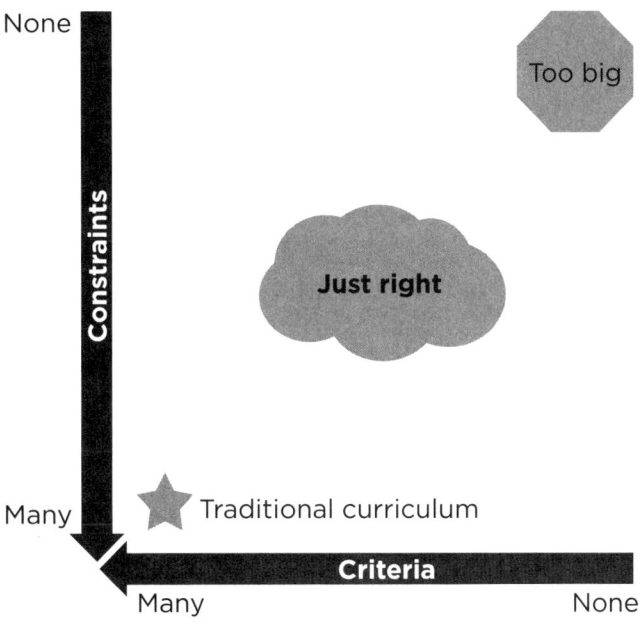

Figure 2.7: The design sweet spot.

These constraints and criteria give your students autonomy from the beginning of a project, which increases engagement and intrinsic motivation (Ryan & Deci, 2018). Giving them a space, rather than a point, to begin from allows students to tackle the same challenge while following diverse paths. They are more likely to reach a different endpoint if they start out at slightly different points in the initial design space. This reinforces the idea that there are multiple solutions.

How the group defines the problem will lead members to their *unique* solution. The key to successfully helping students manage their engineering design projects is to repeatedly bring their focus back to these early definitions. They should be able to relate their choice of what option to prototype to this design space when they make an initial design plan, and connect design decisions and modifications with constraints and criteria, indicating ways their prototype meets these conditions. The boundaries foster critical thinking and analysis, which should be present in all phases of the process, not just in the initial problem definition. Helping students base decisions on constraints and criteria gives them a way to navigate through choices and options. One of the most effective tips I give teachers when I am in their classrooms as a coach is to bring groups back into focus by referring to the design space and the constraints and criteria that are in place. The following sections will help you and your students develop constraints and criteria.

Develop Constraints

Although constraints are often described as *limitations*, it is sometimes better to think of them as *innovation drivers*. If we never had to deal with issues of time, money, or resources, we would have very little need for new solutions and technologies. In most cases, initial constraints are limits that all design groups in a classroom will have in common. In many cases, you, as the teacher, dictate the constraints by setting the challenge or problem. Constraints based on time, money, and resources surface in most challenges. Many constraints that engineers deal with are also due to basic science (think of materials options, the effects of gravity, environmental impacts, and so on) and external regulations, such as those for health and safety.

> Constraints based on time, money, and resources surface in most challenges. Many constraints that engineers deal with are also due to basic science and external regulations.

Have students review some products they use every day and then list what constraints the designers and engineers might have faced. Using a table similar to the one in figure 2.8, encourage practical thinking for the center column and crazy ideas for the right column. By giving the students the option to imagine the impact of not having that constraint, you do two things: (1) you make it more apparent that the given constraints really do limit the possibilities while highlighting the idea that new, innovative designs will have to overcome that constraint if they hope to be adopted, and (2) give students a chance to do a little divergent thinking, which is the focus of phase 2. Removing a constraint is often a great paradigm shift when idea production seems to slow down during brainstorming. Figure 2.8 features a few car-related examples of student-indicated constraints.

Constraint	Design Feature It Led To	If There Were No Constraint
Cost	Simple materials that are strong and easy to find	A car made entirely of titanium because it's so sturdy with fleece-lined seats for comfort
Size	General-sized car that holds between four and six passengers; fits on roads and in garages	A classroom with wheels; big enough for twenty people and a snack bar
Need for fuel onboard the car	Gasoline or diesel were first commonly available; led a need to manage weight and location of the engine	Super lightweight car that connects to a roadway that pulls it along

Figure 2.8: Identifying constraints and their impact example.

Visit **go.SolutionTree.com/21stcenturyskills** *for a free reproducible version of this figure.*

Develop Criteria

While constraints may be the same across all groups developing a solution to a challenge, criteria are most likely different. *Criteria* are the goals a group defines as being the hallmarks of a successful solution. Criteria form each group's identity and should also reflect how members understand the targeted end users' needs.

Much of what we think of as brand identification comes from designs intended to reflect corporate criteria. Let's return to our earlier example of a car. There are so many car models available. In order to function, be safe, and appeal economically to the average consumer, all cars are designed within many of those same constraints. The differences in body styles and interior details are all designed to address the criteria that matter to specific users while reflecting each particular company's brand image and technical strengths. Criteria and a focus on the end user play a large part in making a Hyundai different from a Mercedes, even though both cars obey the laws of physics and accomplish the same goal—providing safe transportation for their customers.

Keep in mind when student groups develop criteria, they are providing information that should form part of the rubric for assessing a successful design. Students should be able to indicate how a solution meets their most important criteria and how the given constraints impact the solution.

Once students identify some criteria, they must determine an order of importance to help guide their design decisions. For instance, suppose two groups are designing a carry-on suitcase. If one group feels low weight is more important than lots of features and durability, and the other group has the order of importance just the opposite, their bags will look very different. Although both groups have the same criteria, both value them differently. Students should reflect this difference in their final design. Students can rank, even rank and weigh, their criteria. Ranking alone is generally fine, particularly for students in grades 3–8 or even less-experienced high school engineers. Weighting criteria involves more calculations (and time) to normalize ranking values. Chapter 4 (page 95) provides a criteria-ranking activity and some ideas to organize the process.

Take a Big-Picture Look With Students

Now that students have input for mapping the design space, it is time to help them take a big-picture look at where they are starting from and roughly where they hope to go. This is not about drilling down to details; encourage students to look at where things fit in terms of information, ideas, and possibilities. Can they envision a way forward or are there still gaps that need to be filled in? To successfully develop a solution to the engineering design challenge, there are some things students will need to know.

Students should think of this information as falling into the following three areas.

1. Background curricular concepts
2. Information about the end user, including how, where, and why the end user might use a solution
3. What has been tried before

Students discussed their end user earlier but some new concerns may have arisen as they identified constraints and criteria. In addition, students have probably started realizing that they need some concepts related to content. Involve students in an initial discussion of what they think they need to know to get started. Involving students at this point reinforces critical thinking and the idea that they can have some ownership of their learning. Have students complete the form in figure 2.8 as a whole class or in groups. You can read more about groups on pages 70–78.

In many projects, groups have been determined after an initial hands-on activity and a very general hook-type video, TED Talk, or article. Once students reach this first phase of the EDP, they should be in groups.

Figure 2.9 (page 40) is an example of a chart for elementary grades based on a project that challenged students to engineer a way for polar bears to stay cool (developed for grades K–2). This form can be used at all grade levels. Teachers for grades K–2 should complete the form with the whole class together. Grades 3–5 can either complete in groups or in a class discussion; middle and high school students can work in groups. Visit **go.SolutionTree.com/21stcenturyskills** for a free form to use with middle school and high school students.

Name:	**Project:** Stay Cool, Polar Bear!
Directions: Please write down any ideas that you have about the following.	
What topics have we talked about?	Climate change and melting ice How we stay warm and cool How the polar bears live and eat
Who is the end user?	A polar bear
How will the end user use the solution?	He will get too warm and may need a way to stay cool.
Where will the end user use the solution?	In the Arctic
Why will the end user use the solution?	It might be too hot for him to stay comfortable.
What has someone tried before?	We make our houses and clothes to help us stay cool when it is too hot.
What else do you think you need to know about?	What materials help us stay cool? What kind of houses do polar bears like? How does the sun make things warm?

Figure 2.9: Elementary school students decide what they need to know.

*Visit **go.SolutionTree.com/21stcenturyskills** for a free reproducible version of this figure.*

One of the most frequent questions I hear from teachers is "How much should I teach, and what can I let students discover on their own?" The decision always comes down to what is *teacher given* versus *student driven*. Sorting this out takes some experience, and it may take a few projects to get it right. And, of course, it will vary from project to project.

Involving students in an initial discussion about how well they understand the end users' needs and helping them identify any concepts they need to understand gives students an opportunity to think critically and to take some ownership of their learning, which is important. In some cases, you may have more insider knowledge of needed content and concepts and this discussion can become more guided

You will need to assess whether students correctly identify the needed curricular content to move forward. Do not go too far into the details, but do make sure students make some sense of needed details once they start looking for them. For example, if middle school students are engineering an educational board game to help younger students learn Spanish, you could ask them to summarize what they knew about Spanish when they were eight or nine years old. Their recall may need some adjustment, but it does not need a fully scripted and detailed lesson. Having them characterize their end users is also helpful, but they may need a "visit" to third grade to make more accurate observations. And walking them through the key steps in most games and their instructions is generally enough to get them to understand that writing instructions can be a real challenge. If there are key ideas and concepts that you think are critical to building a framework for knowledge, a short quiz taken by individual students works well after some instruction.

I have watched a teacher sit back in amazement as her students went into detailed discussions of the functions of the structures in the human hand while pitching their prosthetic hand prototypes. It was far more than the teacher would have covered in her honors-level anatomy class, and it was clear students really knew what they were talking about. In short, they developed a level of expertise because they had to in order to get their devices working properly. Students move closer to concept mastery because they have to connect and apply the ideas.

> Students move closer to concept mastery because they have to connect and apply the ideas.

Know Your Options

The next phase of the EDP has students transitioning from the convergent thinking needed to define a problem and create a design space to the divergent thinking to investigate, develop, and consider a wide range of options. It ends with bringing all the pieces together and choosing which solution to prototype by seeing what fits best in the design space.

This stage becomes more challenging as students get older (from about grade 3 and up) and they are more programmed to go from a problem directly to one specific solution. Creativity exercises from chapter 4 (page 95) can help students develop better divergent thinking skills. Always be sure to have students document brainstorming either by writing on and collecting sticky notes, taking a picture of chart paper with ideas, or creating a written summary. This phase is where the real innovation happens, and it should be documented. Don't put this record aside; encourage students to return to this drawing board whenever they feel they need to make significant adjustments to their design.

During this phase, you and your students will research to learn more, brainstorm to go big, and decide on one prototype.

Step Four: Research to Learn More

You can't possibly teach your students everything they will need to know in order to explore a problem and develop a solution. They will often need to add to their background knowledge at this point. The framework that they need to do so is hopefully in place from some of the information you gave them and that they discovered during phase 1.

Think in terms of the three areas of knowledge discussed in the previous section: (1) curricular content, (2) end user information, and (3) previous solutions. As I pointed out, some of the curricular content may need to take the form of direct instruction. It is important to keep your goal in mind—to provide a clear framework students can use to make sense of more specific ideas.

Anticipate that you will most likely need to develop some deeper understandings and curate some resources for students. Students can access these resources when they recognize a need to know. There is generally no need to require an actual research project; this is an *engineering* project. Evidence of understanding will come in the form of application and the documentation of design decisions in the engineering notebook.

Some reliable research websites follow, and you can visit **go.SolutionTree.com/21stcenturyskills** for live links to more online resources.

- Khan Academy (www.khanacademy.org) videos cover different topics, including science and engineering and arts and humanities, some of which are searchable by grade.
- National Geographic Kids (https://kids.nationalgeographic.com) has elementary-geared videos about animals.
- NASA (www.nasa.gov/audience/forstudents/index.html) has content divided into grade bands.
- TED-Ed (https://ed.ted.com) videos cover a wide range of topics, including humanities and social sciences.

These resources might also include apps and videos, including the following.

- Concord Consortium (https://learn.concord.org) has STEM resources for all grade levels on a wide range of topics.
- PhET Simulations (https://phet.colorado.edu) has free interactive mathematics and science simulations that engage students with game-like interfaces. Teacher resources, lesson plans, and student worksheets are available with a free teacher account.
- How It's Made (https://bit.ly/2vf5Jds) features videos about the manufacturing and design process behind items ranging from cupcakes and hats to Formula F race cars.

For the other two areas where more information may be necessary—knowing about the end user and what may have already been tried—experts working in groups make research more efficient and is similar to the way real design teams work. For instance, suppose students are investigating ways to make education more accessible in remote regions of Nepal. In both the classroom and in the professional world, a group or team commonly has jobs like those in table 2.1. Students will have different areas of expertise, but the goal is to leverage all background knowledge to develop one solution. The Increasing Synergy With Teamwork (page 70) section has more information about this part of the work.

Table 2.1: Jobs and Research for Classroom or Education Design Project

Job	Research Responsibilities
Product or design engineer	School structures, classroom design, and the best design for classroom materials
Country expert	Customs, lifestyles, and geography
Marketing or communication coordinator	How to truly reach the end user and what types of products and approaches are generally successful
Project manager	Previous solutions, including what has worked and what has not

Keep the research part framed as students *becoming experts*. Use the following three steps.

1. Create groups and then assign jobs. (Creating groups is discussed more in the Increasing Synergy With Teamwork section, page 70.)

2. Each student, in his or her role, compiles bulleted information about important facts, using either the sources from the curated resources or through some research of his or her own. (This can be a homework assignment, and you might use it as a component of an individual's grade.) Students document their individual research notes in the group's engineering notebook (see Documenting the Process With the Engineering Notebook, page 78) or online in a Google doc.

3. Have groups come together during class (generally a full period for middle and high school students and about thirty minutes for those in grades 3–5). Each student shares his or her bulleted information and sources with team members. This sharing process within the team generally covers what students need to know about the end user and the previous solutions.

Please remember, this is a *design* project, not a *research* project. Don't require in-depth research. You risk a loss of engagement, the possibility of a never-ending rabbit hole of information, and overwhelming your students. Plus, much of the information may turn out to be extraneous. The research goal is similar to the one you had for curricular content—start with a framework and fill in more specifics as needed. Your students will become experts in ways they need to become experts to create a successful solution. I see it happen every time a project is well designed.

Step Five: Brainstorm Possible Solutions

Dr. Seuss (1975) was so on target with his advice: "Think left and think right and think low and think high. Oh, the thinks you can think up if only you try" (p. 38). Students often struggle to think in any direction but linearly. Being forewarned is forearmed—brainstorming is the part of the EDP that may challenge you most. It is also what makes design a truly robust process; brainstorming brings creativity and innovation into your classroom practice and culture. Please don't skip it! When you're ready for this part of the design process, see chapter 4 (page 95) for strategies and creativity exercises such as Stars and Stripes (and Dots), Chindogu, and SCAMPER This! activities.

> Don't require in-depth research. The research goal is similar to the one you had for curricular content—start with a framework and fill in more specifics as needed.

I devote a lot of time during workshops to fostering creative and divergent-thinking strategies and related activities because it takes more work and patience than most teachers anticipate to make brainstorming productive. This lack of fluency in coming up with new ideas is increasingly apparent as students move beyond third or fourth grade. Creativity researcher Kyung Hee Kim (2011) analyzed the decline in creativity as measured by Torrance tests and says "Children's ability to produce ideas (Fluency) increased up to third grade and remained static between fourth and fifth grades, and then continuously decreased, which might indicate children become alert to issues like accuracy and appropriateness of their responses when they generate ideas" (page 291). Students are more expert at the convergent thinking educators generally value in school; older students may also be more aware of peer pressure and the need to conform socially.

Creativity and brainstorming need two things from you in your role as a guide: (1) prompts or starting points to spur new ideas and (2) procedures that value new ideas and individual contributions. Students have to set aside the design space and background knowledge to a certain extent. You must be intentional about this. Students need to go beyond the many known options to consider the unlimited possibilities that may lead to a new solution.

As students investigate and define the problem, they may have formed images and ideas about what might work. If they go directly from problem to solution, projects are likely to be based on tinkering—trial and error—rather than on engineering. Using the word *challenge* rather than *problem* as much as possible helps counter the urge to go directly to a solution and tends to support a process focus. Be intentional about informing your students they will return to the *design space* after they explore the *outer space* of endless possibilities.

As a leader, you also need to stress, model, and honor the notion that we must respect all ideas and there are no bad ideas—which is not true of actions. This phase is all about *quantity over quality* simply because you cannot evaluate new ideas until they are out there. Lots of "crazy" is good here!

Stress that there are no bad ideas and that all ideas need to be respected. Many teachers I work with say or post phrases and prompts such as these during brainstorming sessions.

- Show respect, respect, respect.
- This is a no-judgment zone.
- There are no bad ideas.
- Focus on quantity over quality.
- No *buts* or *ifs* are allowed.

Brainstorms can easily slow down to "brain drizzles," and multiple viewpoints can easily give way to groupthink. Making creativity and divergent thinking part of your classroom culture really takes some practice. Try some warm-up creativity exercises or team-building activities throughout the year. Chapter 4 provides activities that encourage creativity and divergent thinking, as well as suggestions for managing productive brainstorming sessions on projects (pages 115 and 116); a "Brainstorming Summary" reproducible (page 215) in appendix C helps groups sort out their thinking.

Develop a Solution

During this phase, your students will decide on and create a prototype to visualize and analyze, design an experiment for testing, modify to optimize, and share the news. This is generally the longest phase in the EDP, but it is also the one that students are eager to reach. From your perspective, it is often the most chaotic, since different groups are creating different prototypes and students in each group should be doing different things. Getting into the habit of posting key tasks every day can be helpful. Prompt students, in terms of their job roles, about what they should be working on and what they should complete. Some teachers cue this based on engineering notebook forms (pages 213–225), which can be helpful for monitoring how on task groups, and even individuals, are.

Your role during this phase will generally be more hands off. Little direct instruction is needed, and you can circulate and conference informally with groups. Many teachers report that this is where they start seeing and appreciating the many talents in their classroom.

I offer these tips for managing this controlled chaos.

- Have clean-up rules. Be sure students know that the final five minutes of class are for cleaning up materials and finalizing documentation. Have storage available for their group-specific materials and for general classroom materials: bins, shelves, and storage boxes work well. Pay attention to the time and signal when clean-up starts (or appoint a student to do this each week).

- If you need to talk with the class about something (such as a concept that needs clarification or a rule that needs to be followed better), do it at the start of class, before any building starts. If something comes up mid-class, have a predetermined signal that gets everyone's attention, so that you only have to explain things once.

Step Six: Decide on a Prototype

After your students define the problem and explore possible options, they are ready to transition to the prototyping phase. This step is essentially a transition step between phase 2, know your options, and phase 3, develop a solution. Think of this step as a way to prepare to land back on Earth (the design space) after exploring the outer space of endless possibilities. Hopefully, there are lots of great ideas, but the group should only prototype one. This is the time for some democracy and to revisit the design space boundaries.

> Be sure students know that the final five minutes of class are for cleaning up materials and finalizing documentation.

By this point, the groups have developed or received a list of constraints. Any proposed solution will have to meet those constraints, so that should be their first stop. This is typically either a quantitative judgment or a *yes* or *no* answer, and it should be pretty straightforward. For example, "Does the new game cost more than $20?" or "Can that chair hold 200 pounds?" Once they evaluate possible solutions considering constraints, there may still be several options left. At that point, the group compares options to criteria. Remember that groups have generated most of their own criteria. They will often refer to it in qualitative terms, such as *visually appealing, colorful,* or *easy to understand.*

Making generally qualitative decisions based on criteria can be tricky for groups. It may result in a dominant student taking over, or a quiet student opting out of decision making. Because this entire process mimics real-world design teams and many aspects of authentic problem solving, it is highly valuable in developing skills that go beyond the classroom. We have all been in group situations that require some decision making. The ability to move from subjective to objective choices and to make qualitative and quantitative decisions are things we do every day. These are important skills.

Have students select their top three to five designs and then rank designs by voting. This allows individuals, as well as the group, to assess the design most suitable for meeting their agreed-on top three criteria. It helps if students sketch their best options, list some key features, and name each model. They can then fill out a form like the "Design Ranking" reproducible (page 216) and decide how well each option meets each criterion.

When they complete the form, students list each criterion for a model and rank it between 1–5 according to its ability to meet the criterion. Students should vote separately (and without talking to each other) and then total the ratings if you are concerned about their influencing each other too much. If a group seems to involve every member in its decision making, they could rate and decide as a group. In most cases, silent voting and then tallying is a more democratic approach. (You can see the criteria ranking activity on page 108; rating designs can follow the same silent voting process.)

Students should always determine and rank criteria *before* using this form. They consider and rank designs only if the designs can reasonably meet the group's agreed-on criteria.

Step Seven: Create a Prototype

The initial prototype is the first step in the process of developing an optimized solution; it gives you something to work with. Once students decide what to prototype, it is helpful for them to understand the intent of the prototype.

It is very important to stress to students that a prototype is simply a way to help them *explain and test* their intended solution. It is primarily a visualization that helps them make their ideas concrete. It may have some functionality, allowing students to both explain how it works and test to see if it works. But, most likely, the prototype will not be a fully functional device or process. Using some of the activities in chapter 4 that require building will help reinforce this before you embark on longer-term projects. If you require a certain scale, make that clear and make sure students understand that is part of the requirements for a successful design.

Generally, prototypes at this level are also not full-scale models. A model of disaster-resilient housing will clearly not be the size of a house. The prototypes' sizes may be close to reality—the Building a Better Box (packaging) project (page 135) is a good example of that—but there will most likely be features not quite to the designated scale. Too much focus on the precise scale makes measurement a big concern for students, which can decrease their creativity and engagement. You can make scale more important as students get older and as they gain more experience with building and prototyping.

> The main reasons for prototypes are to act as visual aids for explanation and to provide some testing or feedback data in order to make improvements.

Stressing the idea of a prototype being a visualization also minimizes the need for expensive or hard-to-find materials. It also makes the idea that a prototype for a process (such as an improved system in a cafeteria) can be a drawing, flowchart, or video more understandable. The Prototype Time section (page 116) explains more about prototyping, including how to keep it simple and inexpensive, and offers two activities.

Just keep in mind that the main reasons for prototypes are to act as visual aids for explanation and to provide some testing or feedback data in order to make improvements. In engineering design, prototypes are part of the process, not the final product.

Step Eight: Develop and Perform Tests on Prototypes

Testing a prototype is a key part in making engineering design an iterative process. Students use the data and feedback from testing their prototype to make modifications to their design. If the prototype was destroyed in testing (such as a boat that sank), students should at least submit a written plan for modifications. Make sure students always justify their modifications after considering constraints, criteria, related concepts that they learned, and test results. See the Always Ask for Justifications and Connections section (page 14) for more help.

If you abandon the process and those connections at this stage, students will modify with a trial-and-error approach, which doesn't require information synthesis. Prototype test results will just become more dots to add to the collection unless students support connections (or justifications) to the next step; better learning occurs if students connect their test result to their design improvements.

In addition to functionality, engineers test for things like reliability, safety, consumer feedback, and appeal. A simple pen might be clicked dozens of times. It may be subjected to various weight loads to ensure it doesn't easily break into harmful small pieces. Consumers may test-drive the pen to see if they like the way it feels, looks, and writes. Table 2.2 summarizes some of the reasons for and types of testing a prototype might need. The first row highlights the performance or characteristic of the prototype that you are hoping to test; the second lists some features of the test; and the third provides examples of prototypes that would fit each category of testing.

Table 2.2: Types of Product and Prototype Testing

	Functionality	Reliability and Safety	End User Feedback and Appeal
What Are You Testing For?	• Device's ability to function as designed • Impact of changing conditions on functionality	• Consistent performance to set standard • Number of cycles device can undergo • Safety measures	• Aesthetics • Ease of use • Documentation and instructions • Suitability for planned use
Key Features of the Test	• Dependent and independent variables • Most closely related to a traditional science experiment	• Repeated trials • Clear standard for acceptance • Often products that people interact with extensively	• Measures unquantifiable aspects of design • Targets end user needs
Examples of Prototypes This Type of Test Could Be Used For	• Water filters • Amount of energy a turbine generates	• Boxes that open and close • Toys	• Clothes or accessories • Instructions • Apps and games

Testing for functionality is probably most closely related to the types of testing scenarios students are familiar with from science classes. For example, in the case of designing a wind turbine, students could test to see how varying a fan's speed (independent variable) generates different amounts of energy (dependent variable). Just as in a science experiment, they would conduct a reasonable number of trials. When testing a game for feedback and appeal, using a Likert scale (which has a 1–5 rating) survey allows students to average responses and decide which issues merit modification.

In testing for reliability or safety, it is often important to have students think about what a reasonable number of repetitions might be and what defines failure.

Have your students think about what they need to learn from their prototypes. What are they hoping to get data about? It should be something that they will be able to modify or adjust. I suggest focusing on between two and four features in their testing. For instance, if they designed some new packaging for potato chips (to tackle the challenge of plastic waste), it makes sense to check on its ability to protect the contents (a crush test) along with its appeal to buyers (feedback). No one wants to buy crushed potato chips regardless of how cool the package is, and people rarely consider buying something in a dull or boring package even if the chips are in pristine condition.

> Have your students think about what they need to learn from their prototypes. What are they hoping to get data about?

In some cases, you may want to state a condition that a prototype must meet and then specify a test or have students design the test with you. I often encourage teachers using a project for the first time to specify the test and provide it to the students when introducing the challenge. It makes that first project a bit more manageable and it gives students a goal.

Once your students have an idea of how they will use testing results, you may want to add the challenge of designing tests or experiments. Again, this models the real world. Someone has to come up with a way to test all sorts of features and design characteristics. For example, devices and testing have been designed to mimic how people sit on smartphones in order to determine if the phones will bend when they are in your back pocket. Keyboards undergo days of repeated keystrokes to determine life and wear. Infamous crash test dummies hopefully come out of car safety tests with minimal injuries.

Designing a reasonable test for a prototype takes some planning and creativity. You can approach testing in the following two ways.

1. If your students are new to prototype testing, you may want to spell out some testing parameters and procedures as part of the initial design challenge.
2. Provide some guidance in terms of a form like figure 2.10 for functional testing or figure 2.11 for feedback testing.

Prototype tests can check key features or performance, but most of the time, it is difficult to test for everything. Typically, designers and engineers test new products incrementally, feature by feature, before testing the whole. It is helpful to have your students work through testing this way. They will need to be selective about *what* they are testing and *why* they are testing.

Having your students design some tests for a common object is a great way to connect them with the designed world. Choose a classroom object they are familiar with. A student chair or desk works well, but you can also try things like pens, backpacks, shoes, or sports equipment. Have different groups develop some possible tests for the object. You can refer to table 2.2 (page 47) and assign different reasons for testing to different groups, or have each group design three tests, one for each type of testing.

They can either share their ideas verbally or by writing a description and procedure for one of their tests. Figures 2.10 and 2.11 can provide good guidance for that. Students can exchange written

Name:	Project:
What are we testing for or trying to find out?	
How many trials are we doing?	
What are the key steps in the procedure?	
What data do we hope to obtain?	
How will we analyze the data?	

Figure 2.10: Student-defined guidelines for prototype function tests.

Visit go.SolutionTree.com/21stcenturyskills for a free reproducible version of this figure.

Team Members	

Project Title:
Directions: Create four or five questions that you will put on a separate form for your testers. Use a 1–5 rating scale, where 1 means "Not at all" (or the worst answer) and 5 means "Absolutely" (or the best answer). Think about what you need to know to make your solution the best it can be before writing your questions. Create your feedback form and make sure you have enough copies of it for all of the testers. Try to leave some space for comments on your form.
Could you follow all the instructions? (This is an example question for a board game.)   1: Not at all   2: A little   3: Maybe   4: Most of them   5: Absolutely
Question 1:   1   2   3   4   5
Question 2:   1   2   3   4   5
Question 3:   1   2   3   4   5
Question 4:   1   2   3   4   5
Question 5:   1   2   3   4   5

Figure 2:11: Student-defined questions for prototype feedback tests.

Visit go.SolutionTree.com/21stcenturyskills for a free reproducible version of this figure.

directions with peers to see if others can follow them. Having to describe things verbally or in writing adds a layer of learning. This is good practice for future projects that involve technical writing, such as directions for board games and instructions for product use and processes.

The form in figure 2.10 (page 49) helps students define necessary guidelines for their prototype tests and focus on what needs to be tested in either a fun stand-alone activity or for the prototype that they create in an engineering design project. Encourage them to think about constraints, criteria, and end user needs when testing. Remind them that the success of their prototype is determined by its ability to meet these conditions. If they designed something that they wanted to be highly visually appealing, they should have a feedback question to assess that. If they were testing to see how far their souped-up paper airplane could fly, is one trial reasonable? Probably not—but are one-hundred trials reasonable? Work with them to find a happy medium.

Procedures need to be clearly written to be consistent. You can often refer middle and high school students to handouts they have received for science labs.

> Assign different reasons for testing to different groups, or have each group design three tests, one for each type of testing.

- Ask them what makes the instructions clear. How can you be sure an entire class will do the same thing? Make sure that they understand that the validity of the results relies on doing the same thing each time.

- Ask them if they are getting the right data and what they are hoping to learn from it. For example, if students have built a cardboard chair and they are testing its overall strength, does information about how much weight the seat can hold tell them about the strength of the back if they are concerned about the joint between the seat and the back? Their test needs to measure what happens when someone leans against the back to determine that.

Testing for consumer feedback is sometimes the most informative procedure. For instance, if your students design an educational game for younger students, they will need feedback on whether the game is fun and informative. They might also want to test things like the clarity of the directions and the overall layout and packaging. This happens all the time in the real world of toy and game manufacturing.

Getting consumer feedback presents an excellent opportunity for students to do some data analysis with Likert scales; this makes students aware of the impact of their *expert blind spot*, a common problem people tend to have when they are deeply involved in a design, a product, or information. In elementary or middle school, it might be fun to develop a scale survey (of no more than six questions) for your classroom, or the school buses, or anything else your students experience daily. Collect the responses. Then show students how to find an average and then decide on next steps for improving the design. Details about testing a board game project are in chapter 6 (page 149).

Step Nine: Modify to Optimize

Things are never done in the world of engineering; they can always be better. If that's true, why do engineers stop and declare a product the final version? This is a great question to ask in order to bring your students back to the design space. Engineers find the best combination of features and ideas to meet both the constraints and the criteria. An ideal solution works as well as possible to meet design space condition (end user, constraints, and criteria) with minimal negative impacts. In other words, the solution is *optimized*.

Practicing modifications might be one of the most important things you can do both for developing students' life skills and really understanding how our designed world comes into being. Unfortunately, it is one of the steps I often see teachers skip. Time is always the culprit—the project has taken longer than expected, the prototype has been made and tested, and it seems like a natural ending. But, it is not the end, since *engineers are never done.*

Students learn to appreciate the fact that their solution does not need to be perfect in the beginning. They will begin to value the iterative process of design. One fifth-grade student explains her "favorite part of this project was building a prototype that didn't work and then remaking the project to be much better than before since we could redo things to be better and more efficient" (A. Gallo, personal communication, April 18, 2018). In the end, knowing modifications are needed and expected gives students the confidence to try and moves them away from the convergent approach to finding that nonexistent one right answer.

Approaching an optimal design is usually a step-by-step process, and the modifications students make after prototype testing move the design closer to an optimized version. With that in mind, try to allow time for modifications after testing, ensure that modifications connect to testing results and data, and allow only one modification at a time. It doesn't sound too hard, but teachers often shortchange or bypass this step in the EDP. It will take some planning and some experience to include modification in a meaningful way.

Allow Time for Modifications After Testing

Reinforce the idea that the prototype does not need to be perfect to test. This is sometimes a challenge. During my work in Singapore, I found that many students were very resistant to the idea that the first round of testing was not for perfect performance. Some students will need to be reminded that testing leads to modification, which eventually leads to optimization. Testing is what gives students the information to make the prototype better. The prototype needs to have a certain level of completeness to make testing worthwhile, but most students can identify that.

Students are often afraid to test anything that might fail because, as educators, we have conditioned them to get things "right." Again, stress the value of the process over the product. It is rare for a designed product or process to succeed the first time around; if it does, the testing parameters or procedures may not be rigorous enough. In most cases, students will be anxious to modify to see if their ideas are correct, and this stage should not take much time. Generally, allowing one class period for modification works. Depending on how they test the prototype, testing the new version can happen in that class or in the following one.

> Generally, allowing one class period for modification works.

Ensure Modifications Connect to Testing Results and Data

Asking students to complete a simple form like the "Design Modification Request" reproducible (page 221) will ensure they justify modifications and connect them to test results, criteria, and constraints. The form asks, "What are you changing?" "Why are you changing it?" and "What result do you expect?" Have a stack of these forms handy for all projects. The forms only need to be a half sheet, and you should allow sketches or verbal descriptions.

The product engineer or project manager needs to get your signature to make the modification. This, by the way, is exactly what happens in real-world manufacturing! You will need to check the justification (the *why*) against test results, constraints, and criteria. Clearly, this can be a much less explicit process for elementary students, but they will generally surprise you with some connections. Creating a simple form that asks younger students to draw a sketch of what needs to be modified (fixed), what they plan to do, and why helps to keep them focused on the process.

Allow Only One Modification at a Time

Students can establish a cause between the change and the result only if they can isolate what changed. They can't do that if more than one thing changes. Another reason you want to require one modification at a time is to make students think before modifying. In line with this, I suggest you limit the total number of modifications. A good rule is three (in total) after testing. If students know they have unlimited attempts (which is unrealistic in a classroom setting in any event), their original design may not be very well thought out. If students can modify at will and with little connection, the entire process is essentially a trial-and-error project from start to finish. They may learn, but it will be difficult for you to manage the project and make assessments. If they go too far beyond three modifications, their entire design starts to appear brand-new, and they often lose the whole idea of the design process.

> Limit the total number of modifications. If students know they have unlimited attempts, their original design may not be very well thought out.

You may not have time for any meaningful modification (and this may occur more often than you'd like). It may also be true that once tested, students can't salvage prototypes for modification without a lot of rebuilding. For instance, if a model house designed to float in floods sinks during testing, it will probably be too water-logged to retest, and rebuilding is time-prohibitive. Encourage students to keep prototypes for possible presentations or discussions later.

Be prepared to take the following steps instead.

1. Allow the group time to discuss what the next two or three steps would be.
2. Have groups document these steps, give a reason why they would take the steps, and indicate what they expect the result to be.

Allowing twenty or thirty minutes for this type of reflection maximizes the learning from following the EDP and highlights the fact that the prototype does not always have to be perfect. Students will get the message that what's really important is having ideas about how to make it better.

Final Piece: Share the News

The project is finally done! Well, not quite.

It only makes sense for engineers to tell people about their solution. In your classroom, that communication comes in the form of sharing with each other, another class's students or teacher, or a community group or expert. Sharing gives the work value and it gives these young experts a chance to shine. Showing what you did helps you own what you learned. Mishel, a third-grade "plant engineer" from Rhode Island put it this way in a post-project interview: "My favorite part about the project is to show what we learned in the end and show how it all worked out—and show that we know that maybe we could become plant experts in college" (B. Colahan, personal communication, June 5, 2019).

The students' ability to share their work deserves to be an important part of your assessment. In fact, their ability to explain the process of design and to speak about why something did not work might be more valuable than showcasing a successful product or process. Take it from Mishel!

This is a good time to bring in your school's technology specialist, art teacher, or media staff. Think of anyone who can relate to their project. If that does not yield any results, a panel of teachers or administrators pretending to be potential investors or end users is often just as good. Let them pretend to be representatives from National Geographic looking for camouflage outfits for photographers, if that was your design challenge. A little role-playing is fun for everyone.

If students know that the audience will be someone other than their teacher or classmates, they will generally perform better. A public product is one of the hallmarks of gold-standard PBL:

> Having a public product ups the stakes for students, leading them to do higher quality work. No one wants to look bad in public. When students just turn in their work to the teacher or make a presentation to the class, they (typically) don't care as much as they do when sharing their work with people from the "real world." (Larmer, 2015)

It can help to know what *not* to do. I have seen some projects totally deflate because teachers require a traditional, highly scripted report and presentation formats. That decision generally comes down to assessment or time pressures. These are real concerns and teachers can't ignore them; they need to plan for them. But try to keep your students' enthusiasm and autonomy high as you move through this last phase. (Assessment is discussed in detail in chapters 4, page 95, and chapter 6, page 149.)

Remember, there is value in having an audience. If you step back and think about it, students generally play to an audience of one—*you*. Teachers are generally the source of feedback and assessment, but this is not much like the real world. Any successful product or effort needs buy-in and approval from multiple stakeholders. Also, students pretty quickly figure out what it takes to gain your approval and that impacts their efforts.

Bringing anyone else—other students, other teachers, staff, administrators, outside experts, potential customers, or parents and guardians—into your classroom for this final communication step helps. A larger audience will generally result in a larger effort. A large audience adds a level of authenticity to the project, with the expert in the field or a targeted end user representing the most authentic experience. If timing and scheduling do not allow for an expanded audience, any sort of ongoing display or online gallery might be helpful for engaging your students more authentically. If only you and your students are present, elicit feedback from the other students, either verbally or via simple online or written forms. If a project has taken multiple days, it clearly deserves multiple viewers.

> There is value in having an audience. A larger audience will generally result in a larger effort.

Keep in mind this last step is truly about developing all kinds of communication skills—including writing, listening, and speaking. Use this step to leverage some of the skills and comfort your students have with technology by using video and media-supported presentations. What follows are some key ideas and suggestions related to different options—reports and analyses, presentations, gallery walks, marketing pitches, and various media. Choose what works for your challenge. Any reasonable form of communication that informs the audience about a solution can work, but you must ensure expectations and required elements are clear to students.

You can find some specific communication deliverables in the context of projects in chapters 5 (page 125) and 6 (page 149).

Reports and Analyses

Avoid formal engineering reports. There is no need for another large document to map the entire process. Instead, have your students keep a *group engineering notebook*. If you follow the suggestions for the engineering notebook and reinforce the need for connections and justifications, you will have a document that chronicles and details the process as it happens. The prototype is evidence of the product, the notebook is evidence of the process, and any reflection you require from groups or individuals should provide further evidence of learning.

This notebook documents the design process in real time, not after the fact. You can use it on an ongoing basis for formative assessment and at the end of the project as a component of summative assessment. An additional, detailed report is redundant and time-consuming for both you and your students.

In some cases, a brief (one- or two-page) reflection and summary from each group member might be appropriate. Some sort of group presentation can generally include these types of final documents. Reflection and summary are often helpful in longer projects or in projects where everyone had different jobs. Always provide a few guiding questions.

Any post-project summary and reflection should include a statement of the challenge and should highlight the process and the prototype development. For example, the form in figure 2.12 works for students in grades 3–8 (with any appropriate grade-level modifications). You can use the items listed as prompts for a less scripted document at the high school level. I often use this form as a group document, but teachers often use it as an individual summative assessment and have students work on it at home, after the project is completed.

Some teachers give students the option of writing a failure analysis. This is more common with high school students and with particularly challenging projects and goals. You can include some questions or prompts like those in figure 2.13 for a failure analysis. For question 2, you can adjust the level of technical detail.

Having this written option lets students know failure, just like success, can be a powerful learning tool. Failure also may force them to revisit and make better sense of some of the background information. Again, this failure analysis should not be a lengthy document. A few pages are fine, and it can be a group or an individual effort.

Presentations

If you opt to have students present their work—via PowerPoint, Prezi, or the like—try to steer them very far away from using overloaded slides, or reading from slides or note cards. Limit the

Names:
Engineering design challenge:
Design Space
Describe the end user.
List the three most significant constraints that impacted your design.
List the three most significant criteria that impacted your design.
Developing Your Prototype
List some of the options you considered.
List three reasons you chose your prototype.
Discuss some of the challenges you encountered when creating your prototype.
Say whether you needed to modify your prototype. If you did, explain why and how you modified it.
Final Thoughts
Say whether you think your prototype provides a good solution and why.
Describe what you learned from following the engineering design process to develop your prototype.
Describe what you have learned about yourself while doing this project.

Figure 2.12: Project summary and reflection form.

*Visit **go.SolutionTree.com/21stcenturyskills** for a free reproducible version of this figure.*

---

1. Describe in detail what you observed at the time of the failure.
2. Explain any possible causes, including materials and procedures, for the failure.
3. Was the prototype test a good assessment of how successful a real product would be?
4. Can you identify any decisions or changes that may have led to the failure?
5. Were all critical constraints and criteria identified? Can you connect the failure to not having met or identifying relevant constraints or criteria?
6. Describe three improvements you can make in terms of design, materials, and procedures to ensure that this type of failure does not happen again.

Figure 2.13: Questions and prompts for written failure analysis.

*Visit **go.SolutionTree.com/21stcenturyskills** for a free reproducible version of this figure.*

number of slides and how many words can be on each slide. Point out to your students that visuals can be highly effective. Most projects can be presented in somewhere between five and eight slides. Those slides should not contain vast amounts of text. Allow them to use bullets and key phrases. Generally, five lines with about six or seven words each can effectively summarize the message.

Students in grades 3–5 can work on a topic for each slide, such as one describing the challenge, one describing the design space (end user, constraints, and criteria), one listing possible options, one describing their prototype and testing results, and one showing the finished product or describing future modifications. Remind students at *every* grade level that what they have to say is what really matters.

Make sure students also know that they all must be active presenters. Props help and having prototypes available is more engaging.

Gallery Walks

A gallery walk can be effective if there is not time for individual group presentations or if the project has been fairly short or did not require a physical prototype. Large conferences offer gallery walk poster sessions to give more people time to present their work. Have your students create an engaging poster-sized trifold display that highlights the three phases of the process and explains the solution. Make sure students clearly display and connect constraints, criteria, and key features.

Place all the posters or trifolds on the classroom walls, on display easels, or along a corridor. I like to stress that, if well done, the poster should be self-explanatory, but some teachers like to have each group appoint a spokesperson to stay with each poster to answer any questions. Attendees can provide feedback and assessment in a variety of ways. One way involves a simple three-part form with positive questions and comments like those in figure 2.14. They go to the teacher when the walk is complete.

Presenter's Name:	**Project:**
I really liked . . .	
Something new I learned is . . .	
I would like to learn more about . . .	

Figure 2.14: Project presentation feedback form.

*Visit **go.SolutionTree.com/21stcenturyskills** for a free reproducible version of this figure.*

Another highly visual approach is to provide students with sticky notes color-coded to represent the types of questions: for example, pink stands for *I really liked*, green for *Something new I learned is*, and yellow for *I would like to learn more about*. Students and other observers write a phrase or note on the appropriate colored sticky note and attach it to the edge of the poster. You may have to set some minimum requirements, such as requiring each student make at least one comment per poster and giving students a set number of each color to use.

Marketing Pitches

Your students create a corporate vision, hold jobs, and may even name their solutions. Why not let them pitch their product? Let the rest of the class be the "sharks" and play *Shark Tank* (Burnett, Spirko, Fuchs, & Carter, 2016). Imitating the television show is generally a favorite with many teachers and students.

Why not let them pitch their product?

You will need to require certain elements in the pitch, such as the following.

- A description of the end user's needs or pain point
- What has been tried before this
- The advantages of the solution
- What is unique about this solution
- The "price" (which is always helpful for consumers!)

Watching the "Scrub Daddy" *Shark Tank* episode (Burnett et al., 2016; www.youtube.com/watch?v=ggi3yfUv0Mo) with your students ahead of time will give you an opportunity to highlight elements of a good pitch. Many teachers like to keep the group's classmates engaged by asking them to evaluate the pitches, either after discussion with members of their own team or individually. A form similar to the one in figure 2.15 works well for the evaluations.

Your team: Team members:			
What team is making the pitch?	**What worked well?**	**What could be better?**	**Would I buy or use it?** 1 = No 2 = Maybe 3 = Yes Choose one
			1 2 3
			1 2 3
			1 2 3
			1 2 3
Remember the following things. • How well does this design meet the criteria? • How well does this design work within the constraints? • Does this design meet the end user's needs? Please be thoughtful and respectful!			

Figure 2.15: Pitch feedback form.

*Visit **go.SolutionTree.com/21stcenturyskills** for a free reproducible version of this figure.*

Media

Students can create videos to act as public service announcements, how-to guides, or infomercials. They can create marketing brochures or pamphlets with technical information. You can have students work in class with their group once they have completed and modified their prototype, or have them work at various times throughout the project. Alternatively, this can be an out-of-class assignment with a one- to two-week deadline. Check with your school's technology teacher to see if students have experience with online presentation tools or if the teacher can provide assistance to students.

I have seen teachers use some of these online resources.

- AsSeenOnTV (www.asseenontvlive.com) has examples of infomercials to share with your students.
- Canva (https://bit.ly/2RxUjK4) is for creating marketing graphics and related documents such as posters, brochures, and flyers.
- Glogster (https://edu.glogster.com) is for creating 3-D multimedia posters.
- LucidPress (www.lucidpress.com) has lots of templates for brochures, flyers, newsletters, and related marketing materials.
- Powtoons (www.powtoon.com/home) is a free video-making tool with lots of examples of video ads.

Going Forward

This chapter intentionally leaves out specific project details in order to give you a chance to focus on the skills-building potential of the EDP. It is worth taking a step back and identifying what steps and practices are similar to processes you already use in your classroom and in your own life. Try to highlight some of those natural engineering tendencies by using some of the activities and ideas in chapter 4. I rarely see success if a teacher new to the idea of using the EDP tries to run with a large-scale project before they have walked through some simple, engaging activities to introduce various parts of the process. Creativity and problem solving have always been at the heart of engineering. Think of the EDP as the way to bring them into the heart of your classroom. That is how students will learn to design the future and create the world that never was.

CHAPTER 3
Designing Projects

If you want to build a ship, don't drum up the men to gather wood, divide the work and give orders. Instead, teach them to yearn for the vast and endless sea.

—Antoine de Saint Exupéry

Have you ever faced the question, Why do I need to know this? The student who asks this question is no different than a boat builder who has never seen the sea. Without a sea to sail and places to go, the boat loses meaning. Supporting student-led learning and engagement requires you to flip that model. Don't ask students to "build the boat" until you have shown them "the wonders of the sea" and the adventures they can have. Take your students out of the classroom and into the real world and then work with them as they learn what they need to know to develop solutions to problems. Show them the value and impact of what they will learn *before* they gather the facts. Give them "the beauty of the sea" before you give them the challenge of navigating it.

This chapter focuses on how to combine the EDP with *your* content to create projects that engage and empower students. This chapter actually engages *you* in the process of developing *your own* engineering design projects and empowers *you* to engineer new learning experiences for students. Summaries of actual project plans and resources are in chapters 5 (page 125) and 6 (page 149). They will be useful resources as you begin planning your own projects.

These project-planning tips and tools apply to any topic at any grade level and work with existing projects. However, many of the specifics in this chapter apply best to grades 3 and up. Read the Best Practices for Elementary Students section (page 126) in chapter 5 to help you manage and document the work of those younger engineers in grades K–2. Choose what works for you and don't be afraid to use the guidelines to modify or create projects that fit your classroom better. Just as there is never one right answer to an engineering challenge, there is never a perfect, one-size-fits-all project. After exchanging ideas with other teachers, I've seen projects for prosthetic hands become prosthetic knees; board games become pop-up cards and 3-D puzzle cubes; traffic modeling become a cafeteria redesign. Sharing projects with other teachers works because all the projects follow the same basic principles of project design and all projects follow the EDP.

Choosing a Challenge

Every project starts with a challenge. The *challenge statement* you give to your students is the destination; the *content* is what you need to teach as part of your curriculum; and the *process* is where the *skills* live. The process provides a road map and helps you manage your biggest constraint: time. The process also lets you identify milestones related to curricular content or project management, and allows you to plan more predictable chunks or tasks. For now, there is no set order. Start with your primary concern.

I have, at different times, led with each in terms of my planning. For instance, if you really want your students to learn about focusing on customer feedback as a form of testing, you would more likely choose a challenge that involves designing a game versus building a bridge. If you are short on time, but really want to expose your students to some aspects of engineering design, you might choose a project that doesn't have a lot of technical content and opt for students to re-engineer something they are familiar with. (The Super Sock challenge I pose to teachers in workshops covers all the steps in the EDP in rapid fashion, as participants develop a marketing pitch for the best socks ever.)

> You can build an engineering design project or challenge around anything if you think through the planning stages. Socks, desks, backpacks, even cafeterias are all things students can relate to, and most need some redesign.

You can build an engineering design project or challenge around anything if you think through the planning stages. Socks, desks, backpacks, even cafeterias are all things students can relate to, and most need some redesign. As you design, make sure students do not need to be content experts as they start. But always keep the "ceilings" high—let them explore challenging options and ideas, which often lead to a more in-depth understanding of the overall topic.

Keep in mind, the challenge sets the tone for the reality that problems and solutions don't come in nice, neat boxes; we must work with them to understand them. The design challenge needs to leave room for creativity and provide enough substance to support meaningful collaboration, so it must be relatively broad. This does not mean, however, that the design challenge should be totally open-ended. Constraints drive creativity; we have to develop new solutions when we face limitations. For example, Dr. Seuss (1960) famously wrote *Green Eggs and Ham* to meet the challenge of a fifty-dollar bet from Bennett Cerf, his editor and founder of Random House. Cerf challenged Dr. Seuss to write a children's book with no more than fifty words. Dr. Seuss won the bet with the publication of his most popular book (Clear, n.d.).

Good design challenges address an end user's pain points (needs) and most have a constraint or two built in. Consider options A and B in figure 3.1. In all cases, option B is the better choice for a range of reasons. First, the challenges are more real-world scenarios. Your students will design for people because the end user appears in the challenge statement. Plus, I think most would agree that option B (in both cases) just seems like more fun. Note that option B statements still have some constraints. For the water filter project, size and available resources create limitations. Option B for the building

Challenge Topic	Option A	Option B
Filtering water	Design a device to filter water.	Design a portable water filter young children can carry to school. It should be made of local resources.
Building a structure	Build a scale model bridge three feet long, eight inches wide, and capable of holding a load of ten pounds.	Build a scale model of a solution that will enable people in the village of ____ (group chooses location) to cross a river fifty feet wide overall and eight feet deep in places.
Welcoming refugee children	Create a teddy bear to welcome young children to a refugee camp.	Design something that will help children ages 3–5 feel more at home in a refugee camp, school, or hospital.
Improving the cafeteria	Design a way for each class to have twenty minutes to purchase lunch and eat it.	Develop a process that ensures that all students have time to enjoy a healthy lunch.

Figure 3.1: Compare challenge options.

project is actually somewhat less constrained. There are size issues, but no specific instruction says students must build a bridge. In the welcoming refugee children scenario, an age range is specified. Option B in the cafeteria challenge asks students to take more of a big-picture look.

Option A for the water filtration project is just too wide open; it is hard to know where to start and will most likely devolve into a science experiment to get water clean versus an engineering project to design a useful water filter. Option A for the building project is overly constrained. Assuming most students have access to similar materials, this option is likely to result in slightly different models of the same basic bridge. The same is true in the welcoming refugee children challenge. Groups will all create teddy bears, and they are likely to look very similar. These challenges limit creativity and hint at limited solutions. The cafeteria challenge involves creating a new procedure and severely limits creativity by treating the issues on a class-by-class basis and carving chunks of time out of what is a larger overall time period. These are all fairly well-defined problems with few of the characteristics of real, ill-defined problems—no messiness there! Remember, the challenge statement that you create to state the project goal should not tell your students what to do. It should give them a problem to solve.

> The challenge statement that you create to state the project goal should not tell your students what to do. It should give them a problem to solve.

Ensuring Content, Skills, and Process in Every Challenge

Your projects should have three components: (1) content, (2) skills, and (3) process. You can make any of the three relatively smaller or larger, but good projects include elements of all three. Follow these three steps.

1. Decide where the project fits in the *content* you teach.
2. Determine what *skills* you hope to highlight.
3. Use the *process* to connect the content and skills, and provide a path from challenge to solution.

The project-planning canvas in figure 3.2 models business-planning approaches; visit **go.Solution Tree.com/21stcenturyskills** for a secondary school example of a disaster-resilient housing project. As you use this template for your projects, think of it as the *sketching phase*. Stick to the basics, and highlight key ideas and concerns. Not meant for detail, this template provides a simple snapshot of the big picture on one page.

Challenge:	
Content:	**Process:** (Timing and Milestones)
Skills:	

Source: Adapted from Osterwalder & Pigneur, 2010.

Figure 3.2: Project-planning canvas example.
Visit go.SolutionTree.com/21stcenturyskills for a free reproducible version of this figure.

The content component is what you already teach or plan to teach; that is the expertise you bring to this project. So far, this book has been about the skills component—using the EDP to bring skills-based learning into your classroom. If you are already a PBL practitioner, think of the EDP as your secret sauce that brings all the pieces of the project together. The learning goals you may already have set very often frame milestones and set the timing for the project process.

Content

Generally, this part of the process seems the most challenging for teachers to plan. Teachers working through their first projects often think they should not teach at all; they want to fully embrace the active learning aspects of PBL. This approach often results in students being unable to make sense of the information they need to develop a solution. Students will always need a starting point; just don't map every detail they can discover themselves.

> At this big-picture stage of planning, you just want to identify the key ideas or topics that relate to your curriculum.

At this big-picture stage of planning, you just want to identify the key ideas or topics that relate to your curriculum. Some teachers do this first and then create a design challenge; others design an interesting challenge and then identify the key content. In most cases, you'll have to do some fine-tuning and go back and forth between the two approaches. In the Building a Structure challenge, for example, concepts related to forces, fluid mechanics (due to the river), local architecture and lifestyle, geographic factors, and economic factors (such as what type of commerce might need to cross the river) could all be parts of background knowledge for the project. If you want to include a focus on mathematics, setting costs for prototype materials can introduce a planning and budgeting component, and the planning stage could include a scaling component.

Too many details stifle creativity and engagement because students will constantly check to see if they got it right. In a sense, too much content can create a highly constrained environment. The laws of physics, realities of cost and time, and style preferences will be there no matter what, and some of the learning may occur through unpredicted discovery.

Samantha Scutieri, a high school mathematics and engineering teacher at Union Catholic High School in Scotch Plains, New Jersey, reports one of her challenges is identifying how much content instruction her students need. It is probably one of the most common challenges and one that might be unavoidable. But as Scutieri notes, it just means you are learning with your students:

> A challenge I feel regularly is that when each project comes around, there's content material or techniques I find I didn't know and am always beating myself up that I should have known its involvement in the project. But I do try talking myself out of the rafters by consoling myself that with these open-ended projects, students are learning in unpredictable directions at times, so we just all learn together. Any teacher involved in project-based learning needs to be comfortable with knowing they are going to learn at the same time the students are. This is a process where we are the guides, the coaches, and there is no position called teacher. (personal communication, August 5, 2018)

Take her cue and realize that you can modify, but there will always be unexpected directions and connections. It is a great learning experience for everyone, not a teaching opportunity for you.

Core Curricular Concepts

Students will need some core curricular concepts related to their challenge. I can almost guarantee they will learn more on their own as they work through the project, but they need a starting point. Identify crucial prior knowledge and a few key ideas related to the curriculum you are focusing on in the project. Remember, you are the content expert and you'll develop this project to teach the core curricular concepts you already need to teach; you don't have to use it to teach totally new content. Your goal is to provide a starting point, then check on student progress to ensure students are making the necessary connections along the way, and help them if they need it.

For instance, if students in your middle school Spanish class are engineering a board game to help grade 3 students learn some of the language, you may need to give your middle schoolers some background about the level of Spanish younger students are capable of learning as well as what might be challenging for that age group, but they may not need much more background content information. Some practice writing and reading instructions might be helpful here too, since most people underestimate the difficulties of technical writing. In a project like this, students do a lot of the background research by interviewing and surveying people about the kinds of games they like.

If you are teaching about the Industrial Revolution in history or social studies, background information on how the assembly line developed may part of your curriculum. If students have the challenge of creating an assembly line to produce widgets, they will learn a lot of the logistics, advantages, and disadvantages as they work through the project.

Asking students in your physics class to design a paper airplane that can fly a certain distance is more likely to become an exercise in trial and error if students have no background knowledge. The simple idea that gravity pulls things down should be enough to elicit ideas about what keeps planes from falling. Identifying *lift* as an upward force, and further noting there needs to be forward motion for it to occur, is probably enough at this point. Students will learn more about wing shape and *Bernoulli's principle* ("as a fluid moves faster, it produces less pressure, and conversely, slower moving fluids produce greater pressure") as they test and modify (National Aeronautics and Space Administration, n.d., p. 4). Additional group instruction to discuss wing shapes in birds and planes can happen at this point if you feel it will help with the design process. Further design and testing will most likely also lead students to investigate and identify air resistance (*drag*) and thrust as being the primary horizontal forces.

Quick Builds and Other Hooks

Before any direct instruction, I find the most effective way to identify and elicit prior knowledge is with an engagement activity. This can be a hands-on quick build, or a short video or TED Talk. Follow up with some discussion, or an investigation, observation, or interview related to any aspect of the challenge. You should always construct engagement activities to raise questions more than provide answers. You want to generate in students a need to know. You are trying to *hook* them.

Quick builds are quick-building or design challenges that intentionally have a hard time constraint of ten to twenty minutes to make failure a real possibility. They are essentially low-risk, high-learning, fun activities. Debriefing a quick build always provides insight into prior understandings while helping students identify what they might need to know more about. For instance, if you try the Paper Tower of Power activity (page 96), have each group test its tower and look for points where the tower bends, moves, falls down, or collapses. Be sure each group watches every other group's testing. Work with students to look for patterns in their different designs and highlight issues where balance (center of mass) may be off, or perhaps their structure has an uneven bottom or bad connections between parts.

The quick build provides a reason for students to learn some more information. How you convey the initial core information depends on your teaching style. I often advocate for a bit of direct, guided instruction (with videos, simulations, and peer discussion) for further idea development. The direct instruction component lets you start everyone with a basic road map. Providing students with a curated list of and access to good learning resources keeps the amount of passive instruction to a manageable minimum.

Skills

An engineering design project allows you to introduce a wide range of skills-based learning into your classroom, but you can't do it all at once. Think back to the discussion of the EDP in chapter

2 (page 27). Different steps highlight different skills. And, different types of challenges allow you to highlight different ways of thinking. For instance, a challenge that focuses on meeting one end user's requirements can stress the need for empathy. My earlier example of the anatomy teacher who asked her students to create case studies for patients in need of prosthetic hands had a strong focus on empathy and developing meaningful criteria based on the end user's lifestyle and interests.

Projects that focus on education and awareness-oriented products have a strong focus on communication. I often use a simple paper airplane project that allows for repeated testing to focus on modification to optimize the design. (The ModiFly activity, page 120, lets you use this in a shorter activity that brings home the idea of modifications.) Projects that start with very messy problems often begin with a strong critical-thinking component as students work to control the mess as they define the problem and identify constraints and criteria. Table 3.1 helps you think through what type of project you would like to try based on the skills you want to focus on.

Table 3.1: Project Skills Focus by Type

Skills	Corresponding EDP Step	Project Type
Critical thinking	Defining the problem (Know your problem phase)	Projects that present a messy challenge, few constraints, and student-generated criteria
	Developing and analyzing testing procedures and results (Develop a solution phase)	Projects that lead to testable prototypes, and projects with significant science content or testable via end-user feedback
Creativity	Considering multiple options and brainstorming (Know your options phase)	Projects that lead to prototypes with strong visual components; typically involving objects that define culture and lifestyle
Communication	Working with a group; sharing final results (Develop a solution phase)	Projects that involve an educational or awareness component, or instructions that lend themselves to a marketing pitch; projects whose end users are an audience not an individual
Collaboration	Developing the prototype (Develop a solution phase)	Projects that are more complex, requiring groups to function well together while managing different tasks; multiple components necessary for either the prototype or the final project deliverables
Empathy	Defining the problem or choosing one option to prototype (Know your problem or know your options phase)	Projects that focus on a clear real or imaginary end user; requires interviewing and observing for problem definition and matching possible options to end user needs

continued ▶

Skills	Corresponding EDP Step	Project Type
Systems thinking	Optimizing (Develop a solution phase)	Projects that lead to a process as the solution allow for positives and negatives analysis, feedback loops, impacts, and the like
Global view	Problem definition or optimizing (Know your problem or develop a solution phase)	Projects that involve understanding other cultures, solutions need to fit culture, resources, and lifestyle

Not every project needs to include every step in the EDP. Keep one or two skills in the forefront.

Again, not every project needs to include every step in the EDP (see figure 2.2, page 30). The school year limits the available time. For instance, if you want your students to spend a lot of time interviewing and observing, they will most likely not have lots of time for extensive testing of their solution. Keep one or two skills in the forefront. Don't ignore the others since they work together to form the EDP, but streamlining some steps is fine and often necessary to keep things moving and to make learning goals clear.

Process

Once you develop a challenge and identify the content and skills you want to focus on, using the same basic EDP steps to frame your projects makes tackling different challenges easier for both you and your students. Students have a way to manage developing a solution to a more complex problem, and you have a way to manage multiple projects at the same time. The only real decisions here in terms of project design should center on the skills you want to highlight while you focus the process on related steps. You probably won't skip steps; some will just get more attention.

Chemistry and engineering teacher Anthony Marmora at Hudson Catholic High School in New Jersey finds using the same process and documenting it in an engineering notebook makes it easier for both him and his students to manage projects:

> Using the engineering design process as the core of all of my engineering units allows my students to become familiar with a central idea to help them stay organized and make connections to everyday applications. All of my projects use forms that are specific to each step of the engineering design process. These forms are always the same for each project, but differ tremendously in responses by groups because each group uses the forms to suit the needs of their own project. My students love how the binder portfolios (engineering notebooks) keep their group organized. (personal communication, August 12, 2018)

The following considerations—identifying key timing for content, setting project milestones, and formatively assessing progress during and after class—will become part of your process and are some of the benefits of using the EDP.

Identify Key Timing for Content

Identifying key insertion points for content will help you avoid packing too much content into the beginning of a project. Multiple studies indicate that students best remember the content they encounter first. Next, they recall what comes last (Averall & Heathcote, 2011; Terry, 2005). The content you present between those two points is what students remember least (Sousa, 2006). This is known as the *primacy-recency* effect. Be conscious of this as you plan any direct instruction throughout the overall project and during individual classes.

Structure a multi-lesson project with beginning and ending times each day to focus on content as needed. Overall, plan to deliver content early in your project and reflect at the very end, with time at the beginning or end of each class to extend ideas or to provide as-needed information.

Use ten to fifteen minutes at the start of a class to elicit prior knowledge and convey basic concepts if you see a need to provide more content knowledge during a project. Letting students work on projects and develop their own additional knowledge is useful during the middle of a class.

The last few minutes of the class might be best for extending key concepts or recapping what students learned.

Set Project Milestones

How do you create a "good mess" and still support a meaningful learning experience? How do you ensure students meet learning goals when the path is not always direct and neat? You will need to have a process with milestones to help you keep the focus on both skills and content.

The EDP has a built-in format that allows you to set these project milestones. You can build these milestones using the EDP steps or on graphic organizers and worksheets in an engineering notebook. For instance, on day three of the project, you might have a reminder that brainstorming should be complete and students should evaluate possible options against constraints and criteria with a goal of having an initial design plan complete by the end of class. The "Project Task Planner" reproducible (page 219) in appendix C can help students with schedules.

Some teachers like to create (or work with students to create) a timeline showing the various tasks that make up a project. You can do this with Excel or the free PowerPoint Timeline extension. Timelines like the one in figure 3.3 (page 68), referred to as *Gantt charts,* provide reminders that tasks are not always sequential and that there are often multiple things happening at once. This brings home the importance of team jobs.

A table with columns listing To Do, Doing, and Done models the *Scrum method*, which various industries use to manage complex projects (Schwaber & Sutherland, 2016). We often use chart paper with tasks on sticky notes to create a kanban board for student groups, like the one shown in figure 3.4 (page 69). Teachers who use this quickly become fans. Giving your students both overall and daily milestones based on the EDP provides a familiar way to turn a messy challenge into organized chaos that follows a common route to different solutions.

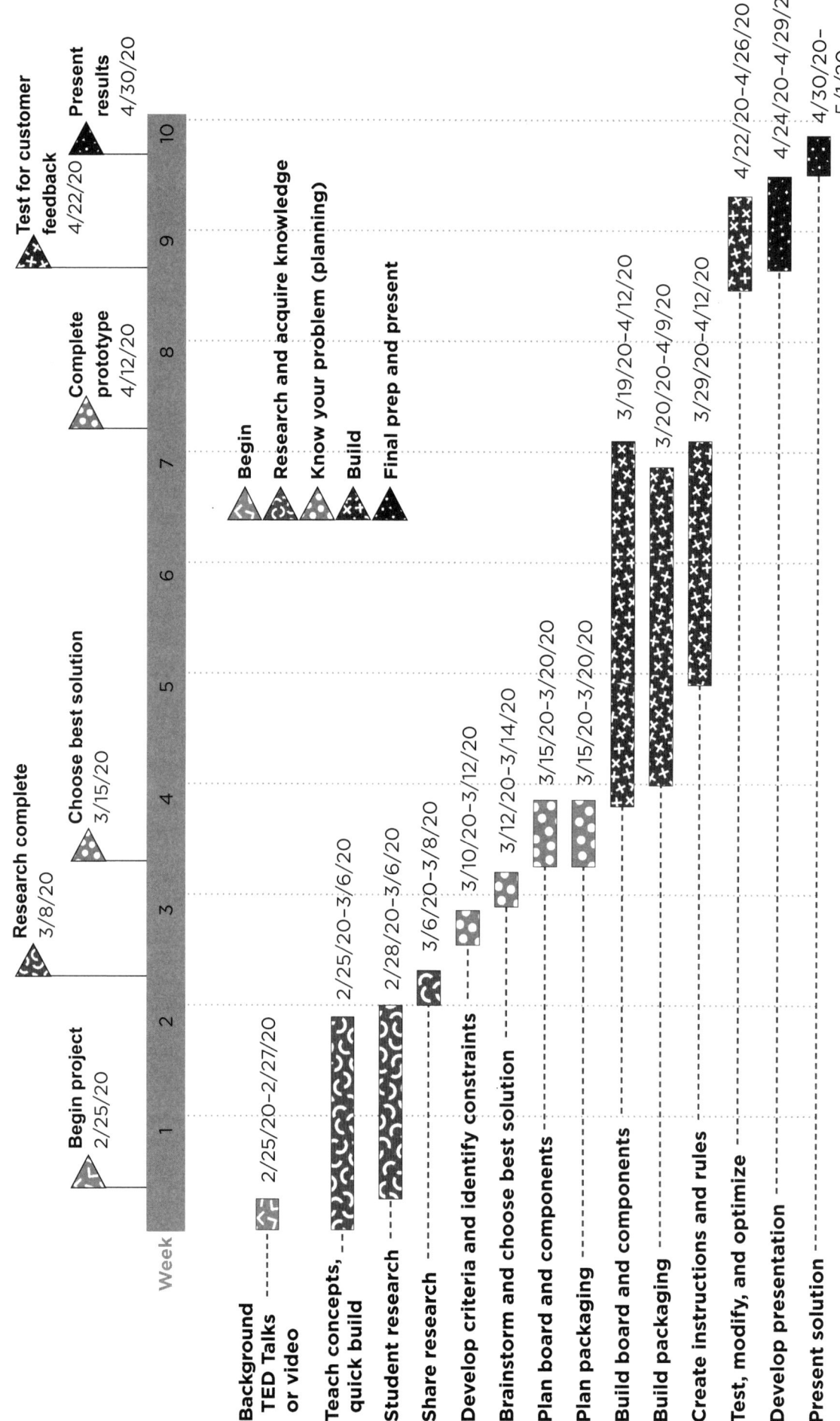

Figure 3.3: Gantt chart showing overlapping tasks.

Figure 3.4: Sample board game project organization—kanban board.

Formatively Assess Progress During and After Class

Students' EDP documentation allows you to formatively assess progress and learning both when students are present and after class. I suggest a group engineering notebook for this. Having a document that chronicles the process each group is following allows you to check in when you have time and when you feel you need to. Not only does an engineering notebook (where students document the process) provide evidence of skills such as critical thinking and collaboration it also provides some insight into content comprehension and proficiency. Students' justifications for design options and features can provide you with information about how well they apply background knowledge and curricular content. You can also use the notebook to identify any misconceptions or areas where you need to teach additional content. If you are using the engineering notebook for grading, make it clear to students when and what you will be looking for. Students should have a chance to revise or complete forms after a discussion with their teacher if documentation is lacking. Your goal is to get them to realize that documenting the EDP is important and useful as they move through the process. Any notebook assessment at this point is preferably formative. It can work as part of the summative assessment at the end of the project since, by then, it should fully and accurately document the process.

Engineering notebooks are discussed in more detail later in this chapter (page 78), and appendix C (page 213) has corresponding forms. You will read more about assessment later in this chapter (page 80).

Increasing Synergy With Teamwork

As you read this section, please keep in mind that students in grades K–2 may not function well in teams because their work tends to parallel each other's. In my experience, it is a good day when you can successfully get these little engineers to work with a partner.

Few designed products or solutions in the real world are the products of one individual. Teams pretty much design, develop, and launch everything. As author and Design Lab founder Don Norman (2013) points out in his book, *The Design of Everyday Things*, "Product development involves an incredible mix of disciplines, from designers to engineers and programmers, manufacturing, packaging, sales, marketing, and service. And more" (p. 238). Norman (2013) goes on to say good design is difficult to achieve and requires collaboration, respect, and cooperation, not just technical skills.

To be authentic, your project should reflect this approach. That means group work. But, poorly managed group work can be painful. However, when it works and the group becomes a team, student learning and growth amplify. Every project I develop involves a design team. I have been in countless classes and seen this work well and not so well. Many of you have a wide range of strategies for managing groups and collaborative learning experiences; use them for these projects.

In addition, the following key ideas can help you manage the tricky landscape of collaboration: choose group sizes carefully, assemble teams carefully, assign different jobs, and involve underrepresented students.

Choose Group Sizes Carefully

Most projects have enough work for four students, so make that your target group size. I don't think I have ever seen a classroom engineering design project require six students per group. If you need to adjust due to class size, think five before three. Groups of four or five typically allow for more diverse opinions and talents. In addition, the impact of one student being absent is really magnified in groups of three. Make sure you have enough work for five students, but try to make the default four.

Assemble Teams Carefully

Try a few short activities or quick builds before assigning students to project teams. Be on the lookout for students who dominate groups, as well as those who sit back. Most of the shorter culture-building activities and quick builds suggested in chapter 4 (page 95) work for groups of between two and four students. Identifying students who do not work well together in a twenty-minute activity can save you and them from a lot of misery in a multi-lesson project. Be really attentive to this when planning your first long-term engineering design project. While we all like to think everyone can learn to get along, trying to manage dysfunctional groups in the context of using the EDP for the first time adds an unnecessary challenge to your attempt to provide a new learning experience. This assembly is a critical step at the beginning of the school year, since you may not know your students and their learning styles and abilities. If you have experience working with groups in your classes, use whatever method you are comfortable with to assign students to work together. Approaches differ, and I have seen a wide range of methods work well.

> Try a few short activities or quick builds before assigning students to project teams.

There are many excellent resources outlining tips, resources, and research on managing groups. Professor Linda C. Hodges (2017) summarizes a wide range of helpful ideas and resources for effective group work at all levels (see https://bit.ly/2JCnDPm). Some highlights follow. You'll see that EDP projects require many of these approaches.

- Tell your students why they're working in groups (Winkelmes, 2013).
- Be ready to handle students' resistance (Aggarwal & O'Brien, 2008).
    - Make sure students see the connection between the project, what matters to them (relevance), and authenticity (real-world work).
    - Minimize your role.
    - Keep groups between three to five students.
    - Divide projects into manageable pieces.
- Help students use good group practices (Johnson, Johnson, & Smith, 1998).
    - Model and insist on active-listening behaviors.
    - Teach students what language to use to express confusion or disagreement.

I always start from the following guidelines, but I have been in classrooms where teachers do the opposite and manage it well.

Pay Attention to Ability Levels

A high school chemistry and engineering teacher with significant experience using engineering design projects in fairly diverse classes says, "Groups that are more varied in terms of academic abilities always work the best. Grouping all of the 'smart' students together often leads to a lot of independent work and very little collaboration" (A. Marmora, personal communication, August 12, 2018). On the other hand, research indicates that "high and medium level ability students benefit more in homogeneous groups but low level ability students benefit more in heterogeneous groups" (Wang, 2013). Clearly, you know your students best. This is a situation where you need to tap into your previous experiences and observations. There are ranges of ability and personalities in each of the mentioned categories.

Keep in mind that your goal is to create the best learning experience for all students. Assuming that the ability of higher-achieving students will rub off on others may not be any more valid than assuming those who do well in school are good collaborators. Most of the teachers I work with find that being attentive to group dynamics in shorter activities and in their early attempts at projects is the best way to develop skills in creating and managing groups.

Avoid All-Boy and All-Girl Groups When Possible

> We don't necessarily work well with our best buddies—even as adults.

Modeling equity in the workplace is one reason for this approach. The other is that research shows mixed groups may learn more than single-sex groups (Gnesdilow, Evenstone, Rutledge, Sullivan, & Puntambekar, 2013). Along the same lines, avoid allowing groups of close friends to work together as well. You may find yourself constantly redirecting them due to their strong social connection. We don't necessarily work well with our best buddies—even as adults.

Avoid Letting Students Select Their Groups Early in the School Year

Students are likely to choose close friends and you will not know them well enough to anticipate issues. Some teachers allow a self-selection method based on the idea of job applicants; students individually identify the jobs they like and then assemble a team from the different job pools. This works best later in the year after completing a few projects, and it takes some moderating from the teacher. It does, however, provide a pretty good real-world model of how design teams assemble.

I have seen a range of group assembly approaches, from random assignment (counting off or pulling names out of a hat) to intentional assignment to student self-selection. I have seen all of these models have a range of results. I know that may not be terribly helpful, but you are less likely to run into group composition problems if you have taken the time to establish culture that supports engineering design and have explained the connection and purpose of an engineering design project to your students.

Working with a group and being part of a team are learning experiences. Not all groups function at optimum levels. Practice what you preach—reflect and modify when you move to the next project. Involving your students in that discussion can be another very valuable teaching moment. Work to get your students to collaborate; it is worth it. Evan and Mishel, two third-graders who engineered a plant for a new planet, put it this way:

> Agreeing with everybody was hard, but the best [thing] was we had all these separate ideas and we had a lot of ideas to come up with a plan. If we had just one idea each, we wouldn't have come up with a big plan. It is not always your way. It is better together. (B. Colahan, personal communication, June 5, 2019).

Address Group Problems

Sometimes, no matter how much thought you put into assembling groups, one or more groups are just not very functional. The How Are We Doing? form in figure 3.5 works well for middle school. Have groups complete it and give it to you after every four classes or so, after you have had an initial short conference with the group, as a checkpoint (and you can be their sounding board). The group should fill in the form together; give them a reminder and a few minutes to do so after those four classes or so. Sometimes being heard and having something to compare progress to is all they need.

Directions: Consider the following reflection statements and rate each for effectiveness.			
	Needs to Improve	**Meets Expectations**	**Goes Above Expectations**
We managed our time well.			
Everyone made an equal effort.			
Everyone listened to everyone's ideas.			
The best group effort of the week was:			
The biggest issue was:			

Figure 3.5: How Are We Doing? group self-evaluation form.

Visit **go.SolutionTree.com/21stcenturyskills** *for a free reproducible version of this figure.*

The "Daily Summary" reproducible (page 220) in appendix C is a way to get a broad look at how groups are progressing.

Assign Different Jobs

Team members need different jobs! I cannot say this enough. There is no reason for everyone to do the same thing. Think of the project as a puzzle where the various pieces come together. Synergy is always more effective than duplication.

Jobs are critical when you are talking about a project that may last several days or be spread out over several weeks. Different jobs give you a way to manage group dynamics and individuals' behavior. They also allow different students to work on different tasks every day—supporting a group multitasking approach is far more productive than everyone doing the same thing or, even worse, sitting around watching one team member work. In almost every

> Different jobs give you a way to manage group dynamics and individuals' behavior.

instance when a teacher tells me groups are not working well, the teacher did not assign jobs or gave students titles but no definite responsibilities.

The idea of design team jobs mimics the real world. Most projects require multiple types of talent and expertise. There are electrical engineers, mechanical engineers, ergonomics experts, visual design professionals, materials engineers, software designers, marketing professionals, and many others involved in designing new models of smartphones. Apple employed eight hundred engineers and other specialists for just the iPhone camera design (Swider, 2015). Obviously, your design teams will be much smaller and the jobs will be much broader!

Specific jobs can vary by project, but there should always be a project manager—but only one! (I have yet to see groups function well with two project managers.) That student should have the best people skills. In addition, think about a job with *engineer* in the title, as well as one focused on design aspects such as visual components, cultural norms, and effective use of materials. Most teams benefit from having a marketing expert as well because how you communicate your product, process, or message to the end user is critical to engaging them. Table 3.2 lists typical jobs that work well for an engineering design project involving developing and building a board game.

Table 3.2: Jobs for Engineering a Board Game

Job Title	Responsibilities
Project Manager	• Starting each class with the group's engineering notebook • Checking in with each team member about the day's focus • Managing the EDP using the forms to document and reflect on progress • Supporting the materials engineer if needed
Design Expert	• Developing the overall game board layout • Managing prototype development and construction with assistance from all team members • Verifying the design meets constraints and criteria • Overseeing modification based on testing • Helping project manager document EDP if needed
Materials Engineer	• Identifying needed materials for the game board, pieces, and packaging • Developing packaging • Establishing the testing procedure • Working with the marketing manager to develop instructions
Marketing Manager	• Creating the commercial communication of the product (such as an instruction manual and user information) • Developing market survey and customer feedback forms • Coordinating with the design engineer • Developing and coordinating a marketing pitch

*Visit **go.SolutionTree.com/21stcenturyskills** for a free reproducible version of this table.*

Be sure that students have these job descriptions. Each group should have a copy in its engineering notebook. You want to use jobs to spread responsibility equally and to introduce the idea of individual expertise. Also, make it clear to your students this is not a work-to-rule environment. Everyone must work together and collaborate to complete the project. Listing responsibilities also makes it clear to students that this project is not just about *making* something and there is no way one person can effectively get everything done. Give students a chance to explore different jobs and roles as you move through different projects. Be attentive to any bias you or they may have in terms of who is skilled at different jobs. The *secretary effect*—giving girls detail-oriented jobs by default—is real (Eveleth, 2015).

Involve Underrepresented Students

It is important to include different populations with different cultural experiences in the engineering career pipeline. Future innovation depends on diversity (National Science Board [NSB], 2018). The NSB (2018) is trying to boost diversity in the engineering workforce by increasing the opportunities for females and other underrepresented minorities (including African Americans, Latinx, Native Americans, and individuals with disabilities) to pursue engineering as a career. We have yet to reach equity in the workplace. If we hope to engage a wide range of creative thinkers in solving our future problems, we need to model it in the classroom.

We need diverse talents, backgrounds, and perspectives if we hope to deal with future challenges:

> Diversity in the workplace not only expands the available talent pool, but also increases the range of perspectives and expertise available to solve challenges. Diversity in the workplace, particularly the STEM workforce, also improves work performance and engagement, enhances the quality of research and provision of health care, and promotes innovation and growth. (National Academies, 2019)

The open-ended, hands-on nature of an engineering design challenge gives all learners a chance to show their skills, talents, and background knowledge; the multifaceted approach to solving problems at the heart of engineering creates a space where everyone can participate and learn from each other. Additionally, you may be wondering how to include students who have physical or language limitations.

The sections on differently abled students and English learners (ELs) address the gaps in representation in the engineering profession and the needs and strengths of those groups in engineering design projects. Finally, the sections on girls and ethnic minorities in engineering reveal workplace gaps and offer resources to help you increase inclusion.

There is no one answer to the challenge of closing the remaining gaps prevalent in a number of STEM fields. I find projects that provide opportunities for collaboration, concept application, and creativity often allow students to develop more confidence. Knowing they can solve a real problem, understanding that failure is a learning tool, and realizing there is often no one right answer are powerful messages for all learners. And learning about career fields that can possibly make our world a better place is a win for every student. As educators, we know firsthand how much representation and models matter: "Kids determine what they can be based on the examples around

> Knowing they can solve a real problem, understanding that failure is a learning tool, and realizing there is often no one right answer are powerful messages for all learners.

them" (Thomas, 2016). Make sure to introduce engineers or other inventors from all countries, races, genders, and physical states to your students. Show pictures of them, read stories about them, and talk about their innovations.

Students Who Have Disabilities

People with disabilities in the United States make up only 1.3 percent of engineers—and less than half a percent of females with disabilities work in engineering and architecture (Bureau of Labor Statistics, 2010). Confirming that people of all abilities can contribute to innovation and engineering is important. Some historical role models include Thomas Edison and Edward Krebs, both of whom were hearing impaired. Temple Grandin, who is autistic, is famous for her work studying animal behavior. Her 2010 TED Talk, "The World Needs All Kinds of Minds" (https://bit.ly/1KCLbNQ) is inspirational for all learners. Albert Einstein and Leonardo da Vinci are believed to have had dyslexia (Helen Arkell Dyslexia Charity, n.d.). Stephen Hawking spent much of his life confined to a wheelchair and unable to speak due to amyotrophic lateral sclerosis (ALS). Hugh Herr, head of the MIT Media Lab and a double amputee, provides amazing firsthand insight into our ability to enable in his 2014 "The New Bionics That Let Us Run, Climb and Dance" TED Talk (https://bit.ly/1qbZd9Y). I have never seen it fail to inspire students of all abilities.

It is difficult to list all the ways to modify every project to include students with various special needs. If you are developing your own engineering design project, plan for those needs by assigning jobs that can be adapted to (or by) the student. Their talents and abilities are unique and projects such as these often provide a terrific platform for them. You also can include projects that focus on universal design and accessibility issues. Kimberly Edginton Bigelow's (2012) research (https://bit.ly/2LzNeHS) explains an engineering design project based on *universal design principles*, which are meant to guide design that provides access for all. Bigelow (2012) offers some useful suggestions, observations, and resources in her paper.

English Learners

Engineering design projects can be a highly effective way to make STEM topics and concepts more accessible to ELs. A National Academies of Sciences, Engineering, and Medicine (NASEM, 2018) report finds educators who work effectively with ELs "are more likely to understand that language is learned through meaningful and active engagement . . . with language in the context of authentic STEM activities and practices" (p. 3).

The extensive report goes on to connect active STEM learning experiences with enhanced STEM meaning making (NASEM, 2018). There is no doubt that a traditional biology course, with its extensive content-specific vocabulary, can be far more challenging to someone already challenged to learn a language. Application, visualization, and collaboration can go a long way in clarifying concepts for an EL. A picture or sketch translates well in all languages, and using sketches in documenting design plans and decisions can support someone who is learning English. In addition, the chance to move any academic subject into a setting that supports conversations with team members and not just lessons full of unfamiliar discipline-specific vocabulary can provide more engagement opportunities for ELs. Sketching and

> A picture or sketch translates well in all languages.

building "can invite participation of students with varying degrees of English proficiency—they can show what they know" (NASEM, 2018, p. 71).

Because EDP projects will be full of context-specific content-specific language, it is a good idea to talk with your school's EL teachers so they can prep students with the necessary language before you begin your unit or project. While language barriers may make it challenging to include ELs in projects that focus on collaboration and communication, it is clearly worth the effort. Everyone benefits from the possibility of early inclusion in the STEM pipeline:

> Opening avenues to success in STEM for the nation's ELs offers a path to improved earning potential, income security, and economic opportunity for these students and their families. At least as important, increasing the diversity of the STEM workforce confers benefits to society as a whole, not simply due to the improved economic circumstances for a substantial segment of society, but also because diversity in the STEM workforce will bring new ideas and new solutions to STEM challenges. (NASEM, 2018, pp. 9–10)

The website ¡Colorín Colorado! (https://bit.ly/2YwQbN6) has several articles and lessons that focus on ELs and Next Generation Science Standards (2013) implementation. It also has links to additional resources with ideas to support ELs in STEM classes.

Girls

I received a degree in engineering in 1979 and was often the only woman in a class or, in my career during the 1980s and 90s, a meeting. While there have been significant improvements in the gender gap across many career fields, the representation of women in many STEM occupations is still far from what it could be. When you delve more deeply into STEM careers, less than 15 percent of all engineers are female and women account for roughly just 25 percent of the computer and mathematical sciences professionals (NSB, 2018). Globally, women are underrepresented in STEM. Only 29 percent of people in science research and development are women; it varies by region, too. Only 19 percent are women in South and West Asia; 48 percent are women in Central Asia; and 32 percent are women in Europe and North America (Foster, 2019). Clearly, we are missing out on the talents and viewpoints of a large portion of our population.

The reasons for this STEM gender gap vary and are subjects of a large body of ongoing research. Studies indicate providing role models, increased exposure to real-world STEM career potential, and support of creativity can help increase engagement of young girls in STEM classes. An Educational Research Center of America (2016) report identifies a lack of STEM confidence contributes to the poor retention of women in STEM at the college and career levels. Work by author Sheryl A. Sorby (2009) and others highlight the impact of struggles with spatial reasoning on the retention of undergraduate women in STEM majors.

The National Girls Collaborative Project does a wonderful job curating, hosting, and developing resources to encourage girls in STEM. Its website (https://ngcproject.org/engaging-girls-in-stem) is worth a visit if you need ideas, data, or research to support your efforts. The National Academy of Engineering offers Engineer Girl (www.engineergirl.org). PBS (https://to.pbs.org/30507h9) has a lesson and resources focusing on women STEM trailblazers and how they overcame obstacles.

Students Who Are an Ethnic Minority

The population in most classrooms is becoming more diverse, but those changes are not reflected in the STEM workforce. Role models are important and always helpful (Boboltz & Yam, 2017). Seek out the resources at the following websites for more and share what you learn with students. The National Alliance for Partnerships in Equity (https://bit.ly/2JqgP3M) provides a list of famous African American women in STEM. Remezcla (https://bit.ly/2xpM35B) provides a list of Latinas working in STEM. The National Action Council for Minorities in Engineering (https://bit.ly/2XIMMNY) offers a range of resources for middle and high school students as well.

Documenting the Process With the Engineering Notebook

When someone asks me to identify the cornerstone of all of the projects I develop, I point to the engineering notebook. Remember, the process matters more than the final product. The EDP embeds all the transferrable skills and many of the curricular content connections, and the engineering notebook is your students' best bet at documenting that process. If you make the project all about the product, students will follow a trial-and-error model with a sole focus on what they produce. The toughest part of focusing on the process is making it visible. Think about this daily. Reinforce this by having each group keep and record something daily in in their notebook.

> The toughest part of focusing on the process is making it visible.

Providing students with to-do lists and goals every day keeps the process in the forefront. You can have a list or some prompts up on the board or refer them to a kanban board if they have created one. If they are not using a kanban board, students can work with you to create a list of steps (see the "Project Task Planner" reproducible, page 219) and record who is responsible, by job description, for completing each. They could also create a timeline like the one in figure 3.3 (page 68).

For the first project, once students have their design challenge and you have worked with them to better define the problem, it is helpful to give them some time to plan—again, with your help. Some teachers do this before brainstorming, but for a first project, I suggest waiting until students have settled on an initial design plan for their prototype and then have them plan the steps in the building, testing, and modifying phase. Once students are familiar with the EDP, they can begin the planning process earlier.

Moving throughout your classroom to monitor on-task efforts and discussions gives you an everyday checkpoint. Documenting the overall process is easiest using the group's engineering notebook. This notebook has impressive effectiveness in not only documenting the design process but also providing a means of both formative and summative assessment; it is a cornerstone of ProjectEngin's project and curriculum design. The notebook can include many forms, some of which appear as reproducibles in appendix C (page 213).

Obviously, there are grade-level issues related to how extensive any documentation can be, but even students in primary grades benefit from seeing evidence. I should note that most teachers find having each group record on a hard copy is the most effective way to keep the process in the forefront. Some teachers use forms on Google Docs (https://docs.google.com) or other collaborative platforms. In cases where students are comfortable working online, this works well. You do need to see who is contributing what, and what changes each student is making.

The notebook is a working document and should be a record of all EDP steps. It does not need to be pristine; in most cases, this notebook is actually somewhat messy. You want students to be comfortable making notes and recording what they do in real time. They do not need to type it, and it does not need to be in complete sentences. I generally discourage erasing and suggest that students just cross things out if needed so that they have a record that they can go back to. I also suggest that the teacher keep some extra copies of blank forms around, since some students like to have a neater document at the end, and because you may need one to be able to read what it says! Just remind them to attach the original to the new copy. It helps to remind them that this is work in progress and that it is hard to get anything wrong. Missing work is the bigger offense. Anything related to their work on the project belongs in a binder labeled Engineering Notebook. The notebook should always be present in some form when groups are working on their projects.

Consider the following thoughts when using engineering notebooks.

- **Do not expect students in grades K–2 to manage extensive documentation:** Kindergarten and first-grade students generally just have a folder or packet with three sketches: (1) plan, (2) build, and (3) test. In kindergarten, usually each student completes his or her own sheet, since agreeing on a drawing is too challenging. As students get used to projects and gain some literacy skills, you can expand on that. In many cases, research is the information you share with them or a book they have read with you.

- **It is a group document:** Based on their job descriptions, different team members should have responsibility for different forms and documents. The project manager should ensure everything is there in a timely manner.

- **If it is in hard-copy form, the notebook probably should not leave your classroom:** There is a little-known health law that guarantees if Johnny takes it home, he will be sick—sometimes for days. There also seems to be a similar law related to entropy and multiple universes that says if it goes into a locker, it will come out in pieces or it will disappear into another dimension.

- **Using a half-inch binder seems to work really well:** You can then reuse them from project to project. If you get the type with a clear plastic sleeve on the cover, teams can create logos and put their names on it. A binder also lets you use plastic sheet protectors if you would like; just be sure to keep a three-hole punch handy in your classroom.

- **You can have students work with electronic versions of engineering notebook documents:** Use whatever site you normally use for collaboration. In some cases, this works well; in others, teachers find keeping an updated hard copy in the classroom helps groups manage progress and refer back to previous steps and background information. This is one you will have to work out for yourself and your students. A lot depends on your comfort level with working online. Just remember, whatever the format, the engineering notebook documents the process, which is never neat and tidy, and always more important than the product.

- **It is always a good idea to have a checklist of engineering notebook contents at the very front of the document:** The "Engineering Notebook Checklist" reproducible (page 214) is an example of this kind of checklist. It helps students stay on track. Also, a space where they record their initials to indicate a form related to their job is complete highlights the individual roles that create team success.

Assessing

It's time to tackle that pesky elephant in the room—*assessment*. Designing assessments that truly reflect and monitor learning is always a challenge and it becomes even more challenging when you are assessing projects. In addition, different schools and districts often have different standards and formats for assessment. This section provides a few guidelines about *what, when,* and *who* (individual versus group) to accurately assess student progress and learning. Good assessment is like learning—a work in progress, and there is always room to modify it to meet your and your students' needs.

Engineering design project assessments should value the process over the product. They should also look for evidence of a student's content understanding and skills mastery. Assessing skills such as creativity, collaboration, critical thinking, and communication should be rubric based; all of these skills are teachable and learnable. Skills are also generally assessed formatively, although they do make an implicit appearance in any assessment of the EDP.

Helping students see that they are making progress in these key skills will go a long way toward helping them be open to new experiences and recognize new abilities in themselves and others. This is a valuable part of the learning experience in any good project-based learning activity. PBLWorks rubrics (https://bit.ly/2UbWEL7) can help you begin assessing skills at all grade levels.

In a Council of Chief State School Officers report, author and CEO of the Learning Policy Institute Linda Darling-Hammond (2017) notes:

> Assessment strategies can be thought of as existing along a continuum. At one end are the multiple-choice and close-ended items found in today's traditional tests. These items measure recall and recognition, but cannot measure higher level thinking skills or the ability to apply them. . . . At the other end are assessments that require substantial student initiation of designs, ideas, and performances, tapping the planning and work management skills especially needed for college and careers. . . . Along this continuum, the role of the student also changes from passively receiving and responding to external questions at one end of the continuum, to taking increasing initiative for finding and making sense of information, as well [as] determining questions, methods, and strategies for investigation at the other end. (pp. 6–7)

These descriptions should strike you as being very much in line with what you are trying to accomplish by making engineering design projects part of your curriculum.

How you structure your assessments will depend on the project or activity and your school policies. Keep in mind that the project overall is an assessment *of* and *for* learning (Stiggins, 2017). A project *is* a performance task. Use more conventional assessments, such as records of time-on-task, short quizzes regarding content, and exit tickets, as needed for checking project progress or individual accountability and learning, but resist the urge to make them a centerpiece of assessment for any project.

Performance Tasks and Assessments

Project-based learning introduces elements that will impact how you assess both content understanding and skills mastery. Table 3.3 lists the engineering design project characteristics that make assessing those projects a little more challenging. Keep the impacts in mind as you develop assessments.

Table 3.3: Factors When Designing Assessments for Engineering Design Projects

Characteristics of Engineering Design Projects	Impact on Assessment
Active	Ongoing formation of ideas and concepts values ongoing formative assessment and checkpoints. Use the engineering notebook and documentation of the process to check for progress and ability to work as a team as well as individual contributions.
Nonlinear, systems thinking	Assessment of complexity and connections is challenging (Cheng, Ructtinger, Fujii, & Mislevy 2010). Concept maps are good indicators of understanding.
Multiple solutions or no one right answer	Traditional summative assessments are ineffective. The focus must be on process. Part of summative assessment should require students to explain final design in terms of fitting constraints and criteria, being scientifically justifiable, and having a rationale for modifications.
Skills and content rich	Process matters more than product. Rubric-based skills assessment is usually the most effective. You can use traditional content understanding assessments, such as written quizzes as checkpoints or individual assessments.
Group work	Evaluating collaboration can be challenging. Peer and self-assessment summative assessments can be helpful, and ongoing teacher observations can provide formative assessments (Lai, DiCerbo, & Foltz, 2017).
Process oriented	Process documentation is required for both formative and summative assessments. (The engineering notebook works for this.) Use periodic group check-ins, daily observations, and single-point rubrics (for the group) that focus on EDP.
Messy real-world problems	Summative assessment requires a focus on connections and rationales for decisions. Critical thinking, creativity, key-concept understanding, and the ability to justify design decisions are paramount. Rubrics will help.

Performance Assessments

Performance assessments are a natural fit because they require students to do just what they're doing naturally during the EDP—active learning tasks that solve problems by applying what they know. So, in a sense, the project itself is a performance task. Assessment is built into the project steps; don't think of assessment as being a separate task or assignment students need to complete. You just need to know how to look for it and how to categorize and analyze the evidence of learning.

> Assessment is built into the project steps.

Researchers and authors George H. Wood, Linda Darling-Hammond, Monty Neill, and Pat Roschewski (2007) suggest performance assessments:

> Are better tools for showing the extent to which students have developed higher order thinking skills, such as the abilities to analyze, synthesize, and evaluate information. They lead to more student engagement in learning and stronger performance on the kinds of authentic tasks that better resemble what they will need to do in the world outside of school. They also provide richer feedback to teachers, leading to improved learning outcomes for students. (p. 4)

Rubrics

You will undoubtedly need to use rubrics, and you will probably need to customize them to fit your needs. BIE has 21st century skills rubrics (https://bit.ly/2UbWEL7) free to use or modify if you are looking for ideas on how to assess skills such as creativity and collaboration.

You can summatively assess how well students follow the engineering design process by using a rubric that mirrors the process's main steps. Figure 3.6 is the EDP rubric I use most frequently. Use the engineering notebook, student presentations, and your own observations to complete this type of rubric.

Groups or individual students may find a single-point rubric helpful for both formative and self-assessments. A *single-point rubric* has just one skill, learning target, or criterion. A single-point rubric asks students to find evidence that addresses success criteria. Professor Jarene Fluckiger (2010) explains this rubric style helps students set goals, self-assess, and innovate—skills essential to design.

Using a single-point rubric that details proficiency in the various parts of the EDP helps keep the problem-solving process in front of students and validates that the process is a critical component in the successful design of the solution. Figure 3.7 (page 85) is a single-point rubric for middle school students. If you use this periodically throughout the project—no more than three times—point out to your students that they should start by noting what they need to work on in the left column. The far right is for when they go above and beyond. You can also just use this once at the end of a project if you prefer. Some teachers do that and use it to get their students to reflect on their work or incorporate this student component into their summative assessment.

There are rubrics and online rubric creators available for a wide range of topics, disciplines, and assessment goals. One of the most extensive compilations of rubrics and related assessment resources is on educator Kathy Schrock's website (www.schrockguide.net/assessment-and-rubrics.html).

	Exceeding Expectations	Meeting Expectations	Approaching Expectations	Developing
Define the problem. MS-ETS1-1	Shows a clear understanding of the problem to be solved. Rephrases the problem clearly and precisely.	Shows a basic understanding of the problem to be solved. Rephrases the problem clearly.	Shows limited understanding of the problem to be solved. Rephrases the problem with limited clarity.	Lacks understanding of the problem to be solved. Does not rephrase the problem.
Identify constraints and criteria. MS-ETS1-1	Identifies and clearly defines all the criteria. Specifies all the constraints with detail.	Identifies most of the criteria. Specifies most of the constraints.	Identifies minimal criteria. Identifies minimal constraints.	Identifies criteria that are irrelevant. Identifies constraints that are irrelevant.
Generate solutions. MS-ETS1-2	Generates an extensive list of possible solutions and thoroughly documents all ideas (list or diagrams).	Generates several possible solutions and documents ideas (list or diagrams).	Generates a single possible solution and documents the idea.	Generates an idea that is unreasonable or does not document ideas.
Develop a prototype and test. MS-ETS1-3 MS-ETS1-4	Prototype meets the task criteria in insightful ways. The model or prototype is constructed with care, neat, attractive and follows plans accurately.	Pro-otype meets the task criteria. The model or prototype is constructed with care but may be missing details.	Prototype meets the task criteria to a limited extent. The model or prototype is messy or missing details.	Prototype does not meet the task criteria. The model or prototype is incomplete.

Figure 3.6: Engineering design process rubric.

continued ▶

	Exceeding Expectations	Meeting Expectations	Approaching Expectations	Developing
Modify to optimize. MS-ETS1-4	Significant improvements are made to the design based on prototype testing and evaluation. Evidence of modification, testing, and optimization are thoroughly documented.	Some improvements are made to the design based on prototype testing and evaluation. Evidence of modification, testing, and optimization are documented.	Minor improvements are made to the design based on testing and evaluation results. Evidence of modification, testing, or optimization is incomplete.	Improvement based on testing and evaluation is not evident. Evidence of modification, testing, or optimization is missing.
Communicate results.	Provides thorough documentation for all steps of the EDP. Team presents a well thought out solution to the problem and includes a rationale for their solution. Teams shows a clear understanding of the related science concepts and design process.	Provides documentation for all steps of the EDP. Team presents a well thought out solution to the problem. Teams show a basic understanding of the related science concepts and design process.	Provides documentation for some steps of the EDP but does not include all steps. Team presents a single solution to the problem. Teams shows little understanding of the related science concepts and design process.	Provides little documentation for the steps of the EDP. Team presents a solution that does not solve the problem. Teams lacks understanding of the related science concepts and design process.

Source for standard: NGSS Lead States, 2013.

Visit **go.SolutionTree.com/21stcenturyskills** *for a free reproducible version of this figure.*

What Can We Do Better?	Goal	We Got It! (✓)	What Did We Do Really Well?
	Know the Problem: We understand the end users; we interviewed them or researched how they live and the problems they have.		
	Know the Problem: We listed our constraints (limitations).		
	Know the Problem: We decided on our criteria (goals) and gave them a ranking.		
	Know the Options: We researched and thought about what had been tried before.		
	Know the Options: We brainstormed lots of ideas.		
	Develop a Solution: We decided to make a prototype of a solution that fits the criteria and constraints.		
	Develop a Solution: Planning—we made an Initial Design Plan and decided what materials we would need.		
	Develop a Solution: Building—we followed our plan to build a prototype.		
	Develop a Solution: We tested our prototype to see if it would work and we got some feedback from others.		
	Develop a Solution: We modified or made plans to modify our prototype based on our testing or feedback. Our solution is the best it can be (optimized).		
	Develop a Solution: Communication—We presented, pitched, or explained our solution to others.		

Figure 3.7: Single-point rubric for the engineering design process.

Visit go.SolutionTree.com/21stcenturyskills for a free reproducible version of this figure.

Concept Maps

Concept maps provide a visual depiction of interrelated concepts and the ways they connect. The maps "help learners learn, researchers create new knowledge, administrators to better structure and manage organizations, writers to write, and evaluators assess learning" (Novak & Cañas, 2008). Joseph D. Novak and Alberto J. Cañas's work led to a software program called Cmap (http://cmap.ihmc.us/docs/learn.php), which is suited for more extensive mapping. However, the website has some resources about theory and methods of constructing concept maps.

In groups or individually, have students create the maps using the following steps (Carnegie Mellon University Eberly Center, 2016). You can use maps as preassessments or formative assessments during projects or tasks.

1. Ask students a question that takes a broad approach to the content that their project will focus on. For instance, a project focused on flight may be introduced with the question "What do we know about how things fly?"

   Students just write down any idea that comes to their minds regarding flight. They can use sticky notes and call the unorganized list the *parking lot* or *idea bank*.

2. Working in groups, students try to organize the sticky notes by drawing arrows to connect thinking in a hierarchy, from big, general categories to more specific details.

   Remind students that this is not final and that they may need to add concepts or move around. They should also make some notes on the arrows linking concepts to indicate what the connection is. These *linking phrases* can include *is made of*, *causes*, or *is a part of*. Each indicates very different relationships.

Use the maps throughout the project. Start early, as a way to involve your students in identifying what they need to know in the problem definition phase. And have periodic fifteen-minute sessions to allow them time to discuss revisions in their groups and, if needed, have a whole-class discussion.

The maps work for assessing content understanding on both a formative basis, as the project progresses, and a component of summative assessment as the project ends. The maps can be group documents throughout the project or have them become individual records as understanding progresses.

The important things to look for in concept maps are the hierarchy and the linking phrases connecting concepts. Hierarchically, big ideas should come first with details following in the order that makes sense to the student.

Stanford researchers Jim Vanides, Yue Yin, Miki Tomita, and Maria Aracelli Ruiz-Primo (2005) explain easy ways to implement concept maps and suggest ways to assess complexity, content, and connections. Although the article (https://stanford.io/2FO6H3Y) is written for middle school teachers, it is important to remember that concept mapping is useful in all disciplines and that the assessment guidelines are generally the same.

LeighAnn Tomaswick and Jennifer Marcinkiewicz (2018; https://bit.ly/2VRku3Y), at Kent State, provide more detailed information and resources about assessing concept maps. A sample concept map for understanding color and mixing is in figure 3.8. I generally prefer to use sticky notes of

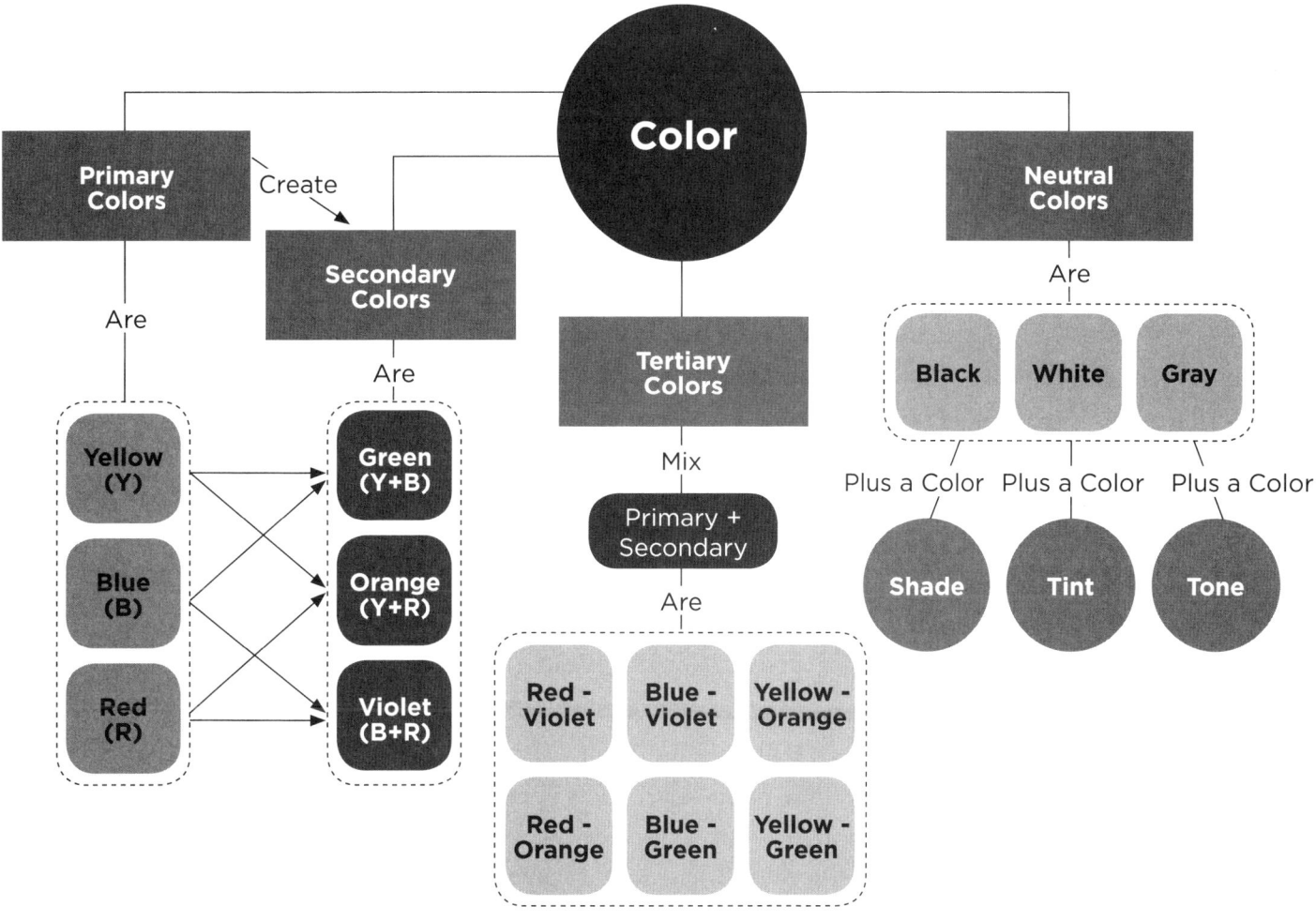

Figure 3.8: Concept map for color mixing.

different sizes and shapes for concept maps. I don't like asking students to fit their thinking into a given framework because that tells little about their understanding. Sticky notes are great because students can rearrange them easily, creating a true picture of the dynamic nature of developing understanding.

What, When, and Who to Assess

Consider these three aspects as you develop and gather student assessments data.

1. Process versus product (prototype)
2. Skills versus content
3. Group versus individual

The *second* part of each pair is a cornerstone for traditional assessments. The *first* part of each pair is where it's especially important to be innovative and intentional. Your assessment will most likely be a combination of both parts.

Process Versus Prototype

In most cases, learning the transferable skills happens during the process and it should, therefore, be a more significant component of assessment. (Chapter 5, page 125, and chapter 6, page 149, note some exceptions.) Students need reinforcement if we want them to realize *how* and *why* they do something may be more important than *what* they do. If the assessment focuses on only the end product or prototype, you risk falling into the following two traps.

1. **Trap one:** The prototype often becomes disengaged from both skills and content learning. It will be very hands-on, but it might not be very minds-on; it might also become a victim of the *bigger, better, faster* mindset.

2. **Trap two:** You will lose many formative assessment opportunities, leaving your students with little basis for reflection and metacognition. The process should count for more than 50 percent of any overall project grade. Moving closer to 70 percent results in better final prototypes because students focus more on following an intentional design process. Trial and error is minimized when you require justifications for decisions and modifications. Requiring documentation and evidence of these justifications supports critical thinking while giving you insight into students' content understanding.

You can include some sort of final presentation or summary as part of the prototype grade if it makes sense for the project. Avoid connecting the product grade to difficult-to-attain performance parameters, such as a paper airplane that needs to fly two hundred feet. While many are comfortable with initial failed attempts and the need to modify (which you want to encourage), a seemingly impossible goal can frustrate them and make them more likely to resort to trial and error. Make the standard attainable but challenging, and consider a bonus of a simple treat, privilege, a few points, or just bragging rights if they go above and beyond. I once had a class that was happy if they got their group's name on the Wall of Fame for impressive performances! They even insisted on making it look like a refrigerator in honor of the way their parents displayed their work at home.

Skills Versus Content

When you are assessing how students use the EDP, you mostly assess skills. Assessment based on evidence of skills is almost always rubric based. Proficiency-based or standards-based grading is an example of this approach, and you can learn more in *Proficiency-Based Assessment: Process, Not Product* (Gobble, Onuscheck, Reibel, & Twadell, 2016). You can also assess specific skills formatively, as discussed earlier, although they do make an appearance in any assessment focused on following the EDP. If you assess how well a group defined a problem, considered the options, and developed and shared a solution, you are assessing evidence of critical thinking, creativity, collaboration, and communication.

You can teach and students can learn creativity, collaboration, critical thinking, and communication. Helping students see they are becoming proficient in these key skills will go a long way in boosting their motivation (NRC, 1994).

Student self-reflection and your own observations are helpful in assessing individual student's skills-based learning. Evidence in the process documentation and product development support assessments of the group's progress toward increasing these skills.

Group Versus Individual

Some schools I have worked with have significant restrictions on any group grading; in others, it is solely at the teacher's or teacher team's discretion.

When determining whether or what to assess and the percentage of a grade for the group versus the individual, consider the following.

- If your assessment matches the project's spirit and goals, base part of the grade on group work. Making all or most assessment components individually focused does very little to promote teamwork and collaboration.

- The group component of assessment should increase in proportion as students get older. Our youngest engineers do not necessarily function well as collaborative teams; they are more like supportive buddies. Encourage collaboration, but do not feel a need to formally assess and include collaboration in the score until students reach at least grade 3. By late middle school or high school, most projects I create have group assessments contributing about 70 to 75 percent of each student's grade. The group assessment generally centers around the overall understanding and implementation of the EDP, the final prototype, and the final presentation. Individual assessments generally center around content understanding and skills development (such as critical thinking, creativity, and collaboration).

- Make sure to still include some individual assessments so your students have some control over their grades. You can employ brief content-focused quizzes or questions, self- and peer assessment, background research requirements, and ongoing observations of contributions and time-on-tasks as individual components of assessment.

- If you suspect most groups in your class might have trouble functioning well, place less emphasis on assessing the group component for the first one or two projects. You want the project to be a success and hope your students will learn much more than you can possibly assess. Minimize assessment issues by understanding that new ways of learning are an adjustment for you and your students. If they feel they have some control, assessment issues shouldn't be a focal point.

Peer Assessment

Consider making it clear to your students that there will be a peer assessment at the end of each project. This lets them know that everyone is equally responsible for successful completion and that there are multiple points of accountability. Most teachers find that the peer assessment is a valuable asset when students start complaining about who is not working in their group or, conversely, to nudge those who seem to take breaks from working.

> Consider making it clear to your students that there will be a peer assessment at the end of each project.

Use the peer assessment results as part of each individual's final grade. It is wise to generally make it a small part (5 to 10 percent), but that is enough to have an impact and it gives students a better sense of control when working in groups.

You can find various peer assessment forms online. The website Teacher-Written Eduware has a number of peer assessment rubrics (www.lapresenter.com/coopevalpacket.pdf) that you might find useful.

Keep the following things in mind when creating a form or finding an existing form to use.

- Do not have each student use a separate form for each team member. It makes it harder for you to compile the information and more difficult to get a good snapshot of a group's dynamics.

- Having a comparable space for self-assessment on the form might make it easier and more insightful to students and to you. *Always* require the student filling in the form to identify themselves. This helps keep things honest and respectful. I have found that students are generally thoughtful and honest when completing the peer assessment. Remind them that it is not their job to reflect on how well someone else understands the content; their job is to assess how well team members contributed.

In my experience, most K–2 teachers use a self-assessment of collaboration or teamwork rather than actual peer assessment. Students in grade 3 can generally manage a peer assessment if the classroom culture supports collaboration and teamwork. PBLWorks (https://bit.ly/2Xkn4LF) has an excellent teamwork rubric for grades K–2; you can modify it for grade 3.

The rubric in figure 3.9 is the one I use most often. It works well for grades 5–12. Try to keep the focus on the ideas represented in the figure if you choose to create your own or modify an existing one.

In conclusion, always reflect on how well your assessments match the project goals and reveal student learning. These are generally the areas where project revision and fine-tuning are most common. Effective assessment free from the right-or-wrong script we are so familiar with can be challenging to develop, implement, and explain. In some cases, this is when a single-point rubric can support a smoother transition as you see assessment as a valuable tool.

I also find it helpful to keep the following two things in mind.

1. The entire project is a performance assessment and, in a way, you just need to chunk it into meaningful pieces. As students follow the EDP, preparation, planning, prototyping, revising, producing the final product, and reviewing can make discrete assessment events.

2. You are not alone in dealing with new ways to assess student learning. You do not need to find discussions specific to engineering design projects. Look for PBL assessment tips, borrow from portfolio critiques in the arts and other fields, and, most importantly, talk to your colleagues and students about what makes a fair and effective assessment.

Name:			Date:			
Group Project:						
Directions: Write the names of your group members in the boxes across the top. For each criterion, rate yourself using the scale. Then rate each of your group members on the same scale. Write any additional comments on the back of the page. 　　　4 = Always　　　3 = Often　　　2 = Sometimes　　　1 = Never						
Criteria	**My Name**	**Student Name**	**Student Name**	**Student Name**	**Student Name**	**Student Name**
Participated in group discussions						
Helped keep the group focused and on task						
Contributed useful ideas						
Respectfully listened to others						
Took responsibility for a fair share of the work						

Source: Adapted from Manis, 2012.

Figure 3.9: Peer group assessment form.

Visit **go.SolutionTree.com/21stcenturyskills** *for a free reproducible version of this figure.*

Going Forward

The ideas in this chapter will help you to create similar projects and those similarities will make project management easier for both you and your students. Mix in the right amount of skills and content and follow the EDP. Bring real-world challenges into your classroom. *What, how,* and *when* it makes sense to adjust and assess will become evident if you design and think in those terms. Think back to the idea of yearning for the sea and boat building. Encourage your students to explore new challenges and possibilities, and they will work to master the tools they need to set sail.

PART II
ACTIVITIES AND PROJECTS

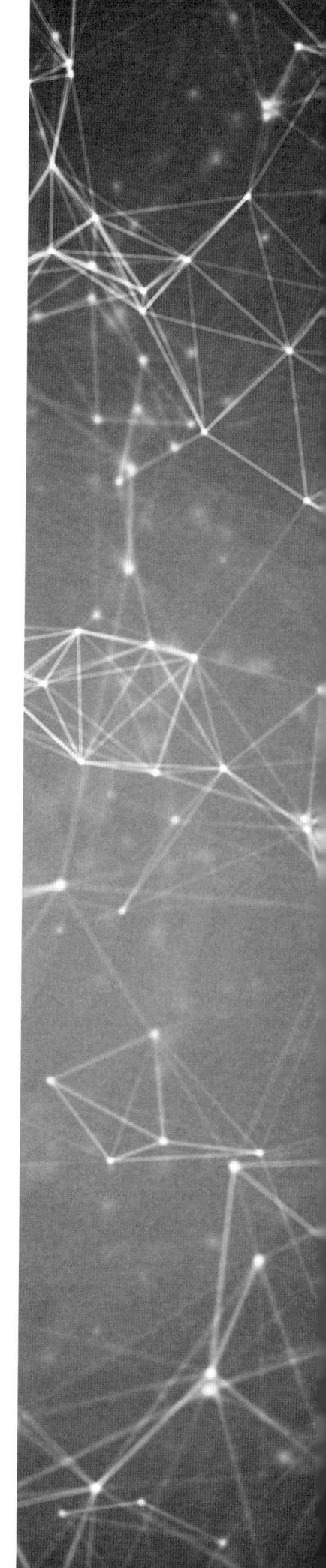

CHAPTER 4
Starting With Activities That Support Engineering Thinking and Skills

You don't learn to walk by following rules. You learn by doing, and by falling over.

—Richard Branson

Reading about the EDP is a good start, but including some of the practices and ways of thinking in your current curriculum is where the real learning begins. If you are convinced of the *why*, it is time to consider the *how*. Blending in small steps and short activities with what you already do can impact your classroom culture and establish practices that will make including engineering design projects a natural next step.

This chapter gives you some ideas for those small steps through activities tied to parts of the EDP and that focus on the skills so important for future learning and employment. These steps provide easy entry points to consider, and they support the culture shifts and hallmarks discussed in chapter 1 (page 7). They are short (generally one class or less) and easily adapted for and implemented in most classrooms. Some may also serve as hooks for the longer engineering design projects in chapters 5 and 6. Throughout this chapter and those that follow, I will use *activities* to refer to these shorter learning experiences and *projects* to refer to longer, multiclass engineering design challenges.

I urge you to try a few activities before jumping into those bigger projects. They let you learn by doing, while keeping the expenses of time, resources, and planning to a minimum. You and your students are starting on a new path. That first full step is often the most difficult, and that first part of the journey should not involve complicated preparations that take students, and you, to remote places. These activities give you a chance

> Try a few activities before jumping into those bigger projects.

to test the waters without fully jumping in. To truly include engineering design in your classroom on the level that combines creative critical thinking with your existing curriculum, you will need to use projects like those outlined in chapters 5 or 6. This chapter is the warm-up to the main event.

These are not full-blown projects that take students multiple class periods to complete. They are quick, low-barrier opportunities for engagement. If students fall over, it is easy to get back up. Most importantly, these activities get high marks as collaborative learning experiences in my workshops and often make it back into participant teachers' classrooms. They will engage your students and provide diverse ways to support what you already do while introducing EDP practices and thinking. Many work in all grades with appropriate modifications for group size (smaller in lower grades), literacy, and available resources (often less in grades K–2). Tips for modifying are noted where applicable. Don't worry about a long process, following lots of steps, or assessment. Use these activities to get an idea of the types of learners, range of talents, and overall social dynamic in your classroom. Try them and have some fun!

The activities are grouped to highlight the hallmark of engineering thinking that they reinforce or the phase of the EDP that they relate to in terms of skills and ways of thinking. Many could easily cross over classifications, so feel free to use them to fit your goals. You're learning how to walk (and falling over), so to speak. Learn to think like an engineer before making the leap into projects.

Learn From Failure

Use some of the following activities to introduce the EDP as a way of thinking and doing separate from any specific challenge or topic. The activities will help you make the engineering design hallmarks and mindsets part of your classroom culture and later you will find moving to more involved projects (necessary for curricular connections) a much easier next step.

The following activities—Paper Tower of Power, Three-Legged Stool, and Flight of the Table Tennis Ball—help students internalize the idea that *failure is always an option.*

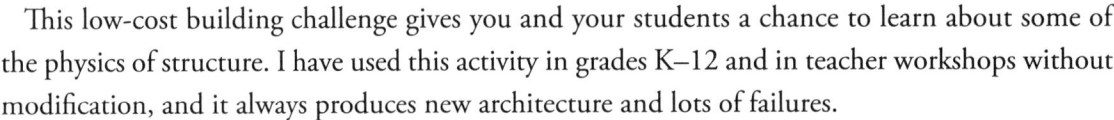

Paper Tower of Power Activity

This low-cost building challenge gives you and your students a chance to learn about some of the physics of structure. I have used this activity in grades K–12 and in teacher workshops without modification, and it always produces new architecture and lots of failures.

Challenge

Students build a tower that can support a tennis ball and resist a hair dryer-force hurricane. You can use this activity as a quick build for the Huff 'n Puff project (page 128).

Group Size

Two to four students

Time

Fifteen minutes (Try to keep to this; it helps ensure less-than-perfect structures and likely failure. You can allow a minute or two more if students do not have standing structures. Note that students in grade K–2 may need twenty minutes; you can also make the tower shorter, maybe twelve inches, so they can easily check height with a ruler.)

Materials

These are the materials.

- Three sheets of newspaper
- Ten inches of transparent adhesive tape
- Scissors

Constraints

These are the constraints.

- The structure must be at least eighteen-inches tall.
- The structure must be freestanding. The tower cannot be supported by students or be taped to the table.
- The tennis ball must sit at the top, not nest inside, the tower. Groups cannot attach the tennis ball to the tower with tape. The ball is the highest part of the tower.

Directions

These are the directions, including students' test parameters.

1. Give each group a tennis ball when it is time to test (after the fifteen-minute building time). Instruct them not to test until it is their turn.
2. The first test (assuming the tower stands) is to put the tennis ball on top of the structure. The ball must sit on top and cannot nest inside the tower or be attached with tape.
3. Instruct groups to observe *how* the tower fails (or handles stress if it doesn't fail).
4. Now try the next test: subject towers to a wind storm if it is able to hold the tennis ball. Leave the tennis ball in place for this test. The teacher should create the windstorm, using a hair dryer, fan, or compressor.

As an extension, you can supply wind from multiple directions.

- **Directly perpendicular to surfaces:** This provides a force that acts like a *push*. This is why structures subjected to wind (bridges) and water (coastal homes) taking a direct impact have open structures for the wind or water to pass through, not push against.
- **Parallel to surfaces:** This provides a force that can generate outward *lift* by creating a high velocity, low-pressure region outside of the building (Bernoulli's principle). This is what causes windows and roofs to blow outward in some wind events.
- **At an angle, parallel to corners:** This creates turbulent flow and eddies due to *edge effects* and can cause a range of problems in structures.

Consider these notes about testing.

- If the testing desk or table surface is very smooth, the low friction will cause a skating effect when the teacher is applying wind. Consider allowing each group to use a piece of felt (approximately twelve-inch square) as the building footprint. In other words, have each group build its tower on the felt surface. If this is not an option, have each group use a little tape to anchor its tower before moving on to this wind phase of testing.

- Having each group test its structure with the other groups observing provides an opportunity for all students to observe the success or failure of lots of different ideas about construction. The opportunity to see patterns and different ways of doing things is a valuable learning experience. I try to make this the rule in most projects or activities unless time or some other issue makes it impractical. Have students discuss how the towers of each group failed or what might make them stronger. Watching multiple tests and failures (and successes) gives the entire class lots of ideas about what might work in a more involved prototype.

- For elementary students, taping ears and whiskers on a hair dryer recreates the wolf from fairy tales.

- If you can adjust air flow, you can pretend to create different levels of storms. Structures in different regions are designed to withstand different categories of tornadoes, hurricanes, typhoons, and so on. Historical data, climate studies, and balancing safety considerations against cost and feasibility often determine the designs.

- If you have a phone or camera available, video recording some of the testing can help students identify causes of failure.

Reflection Questions

You can debrief students with questions like the following.

- Can you identify what first caused the failure? Did it bend, buckle, twist, collapse, or just tip over? These are all different modes of structural failure.

- Was one part really weak? (Teacher note: Many failures have an initiation point and then it spreads due to unloading, crack propagation, and unstable equilibrium. Sometimes all you need to do is fix that initial problem.)

- What might be a good way to fix the problems mentioned in the first two questions?

- Did you notice any common features in buildings that stayed up in the storm or successfully held the tennis ball?

Three-Legged Stool Activity

This is a rapid, fast-paced introduction to furniture design. It works for all age groups. Older students can test to see if the stool can hold weights, rocks, or books; kindergarteners and first graders can try stuffed animals, blocks, or books. This activity is great as a standalone, but it also works well with any project dealing with furniture design including the Just Right project (page 132). Middle and high school teachers can use it as a hook for projects involving the construction of life-sized cardboard chairs and stools.

Do not disclose the testing procedure until all groups are done. When you do test the stools, have each group test while the others watch. Have students discuss how stools failed, as well as what worked and didn't work.

Challenge

Build a stool capable of holding a smartphone's weight. Add to the challenge by seeing how many phones each stool can hold. (Use books if no smartphones are available.)

Group Size

Three or four students

Time

Fifteen minutes

Materials

These are the materials.

- Four sheets of copy paper
- Two index cards, ruled or unruled, 4 × 6 inches
- Twelve inches of transparent adhesive tape
- Scissors
- Weights, rocks, blocks, a smartphone, or textbooks (middle or high school) or a stuffed animal (elementary students)

Constraints

These are the constraints.

- Six-inch minimum height
- Five-inch minimum height per leg
- Six-inch minimum diameter or diagonal seat

Testing

These are the students' test parameters. In all cases, have them start with one object and slowly add to the load until the stool fails. (This is called *testing until failure* and it is sometimes used to find the margin of safety for a consumer product.)

- Students test to see if the stool will hold a book, a smartphone, or a weight. If it holds one, see how many it can hold before failure.
- Alternatively, they can place a small plastic container on the stool and load it with weights, rocks, or blocks until it fails.
- Elementary students place a teddy bear on the stool. Add some weights or a book if more loading is needed.

Reflection Questions

You can debrief students with the following questions.

- What did you think your stool would be able to hold? Were you surprised by the results?
- What started to happen to the legs when the load was too much? What happened to the seat?
- What would make it stronger?
- Did you use any ideas from chairs and stools in the room?
- Do you see any things that the strongest stools had in common?

Flight of the Table Tennis Ball Activity

This activity is best in grades 5–12. A great way to introduce the potential of kinetic energy transfers, this slightly longer activity allows for multiple tests and modifications based on results. This activity is often used to highlight the idea that you can always make a device better through modification. Some teachers like the idea that you can also launch—pardon the pun!—a discussion about precision and accuracy if students mark successive attempts with sticky notes.

It is challenging. I have witnessed only a handful of ping pong balls make it into the cup in numerous classroom and workshop experiences, but even teachers want to keep trying. The energy transfer focus also makes it a good quick build for any energy-based project.

Visit **go.SolutionTree.com/21stcenturyskills** to access free instructions to this activity.

Challenge

Given limited materials, build a device that launches a table tennis ball into a cup in the center of a six-foot-diameter circle.

Group Size

Two to four students

Time

Fifteen minutes

Materials

These are the materials, and every group should have all of these.

- Two inches of transparent adhesive tape
- Twelve inches of three-ply string
- Four rubber bands
- One small paper drinking cup
- One sheet of copy paper
- Two paper clips
- One brown paper lunch bag (approximately 3 × 7 inches)
- Scissors (no glue)

Constraints

These are the constraints.

- Every group member must be actively involved in the ball launching.
- The ball must start outside the circle and come to rest inside a cup in the center of the circle.
- No one can touch the ball or reach into the six-foot circle.
- No part of anyone's body may extend into an imaginary cylinder that extends above the circle.
- Teams may only use the provided materials.

Testing

These are the students' test parameters.

- You may test your device outside of the actual testing area during your building process. But you cannot use the actual testing area.
- You will get three attempts to get your ball into the center cup. Every group member must actively be involved.
- After every attempt, you will mark where the ball landed in the circle with a different color sticky note for each group.
- For version one (quantitative), you must measure and record your attempts from the center cup. In this case, have a tape measure available.
- For version two (qualitative), keep sticky notes in place and look for improvement with subsequent trials.
- After every group makes three attempts, the teacher may allow groups to make one or two modifications and retest their launching devices.

As an extension, consider asking students the following.

- Repeat the challenge with a ball of a different mass. How does this affect the performance of the device? Discuss how this change (modification) further supports Newton's First and Second Laws of Motion.

- Repeat the challenge and give groups any additional materials they request or a selection of materials to add to their devices. How does the modification affect the performance?

- Change the challenge from accuracy to longest distance. Include factors for predicting reproducible results. A Height of Flight challenge has students investigate changing the angle of the launch.

Reflection Questions

You can debrief students with the following questions.

- Can you explain how your device stores potential energy?
- Can you explain how your device transforms energy from potential to kinetic?
- What did the testing measurements tell you about your device?
- Can you identify the forces acting on your ball? Did you make use of any forces?
- Did you need to take into consideration the mass of your ball? How did the overall weight of your ball affect the design of your device?
- What would happen if you use a heavier ball? What modifications would you need to make to your device to accommodate the heavier ball?

Zach Floyd, a fifth-grade teacher at Our Lady of Czestochowa in New Jersey, offers this:

> Using the engineering design process has given me a stronger understanding of what the process can mean to our students. We are providing them with a framework that can be applied to any challenge, interest, or subject they encounter. Students learn to develop empathy for their end user and see the importance of understanding the problem before coming up with solutions. They reflect on their failures and use that knowledge to make iterations and turn them into successes. (personal communication, September 10, 2018)

Know Your Problem

It really is important to define a problem as clearly as possible. The challenge statement is what establishes the starting point for the solution. The Five Ws graphic organizer in figure 2.4 (page 34) is a good place to start. You can weave good problem definition into almost any subject, from identification of causation in history class to the challenges characters in a book face, to everyday life situations, such as the chronically late teacher who keeps hitting that snooze button in chapter 2 (page 27).

For instance, if a business determines it has too much product in inventory, is the problem less customer demand or overproduction? The first problem probably requires a marketing and sales solution and the second problem involves better production planning. These are two very different solutions. Practice identifying problems with your students by asking them to refine problems such as *It takes too long for everyone to buy their lunch in the cafeteria* or *I have too much homework every night.* Are these due to bad planning (too many students in the cafeteria at once) or superb cuisine (everyone wants to buy lunch); resource allotment (too much time playing video games at night) or poor processes elsewhere (too many interruptions in the classroom)?

The other aspects of problem definition that might be new to your students involve identifying an end user's needs and the constraints and criteria that frame the design space. Trying some of the following activities should help students learn how to identify problems.

> You can weave good problem definition into almost any subject, from identification of causation in history class to the challenges characters in a book face, to everyday life situations.

Constraints and Criteria Activity

In chapter 2, I refer to the idea of framing the design space (figure 2.6, page 36) with constraints and criteria to give you that *somewhere* to start. The idea of designing to meet criteria under certain constraints is often an alternate definition of *engineering design*. Chapter 2 discusses some activities and ways of thinking about constraints and criteria, but there are lots of simple activities you can have students do whenever there is unexpected free time in the schedule, or as way to analyze a problem facing a literary character or historical figure.

Since every design starts from constraints and criteria, there must be some behind everything in your classroom. This activity gets students thinking about where designs start. English language arts, history, and foreign language teachers can adapt this approach to analyzing challenges and scenarios characters and people face.

Challenge

Students go on a constraints and criteria scavenger hunt. It can be cooperative or competitive. This works best for grades 3–12 after you have discussed what constraints (limitations) and criteria (goals) are. It may help, particularly with the earlier grades, to first have a whole-class discussion about something in the room, such as a deck or chair.

Group Size

Whole class or groups of two or three students

Time

Twenty or thirty minutes, depending on depth of follow-up discussion

Materials

These are the materials.

- Reverse design space form (see figure 4.1)

Directions

These are the directions.

1. Using a simple graphic organizer like the one in figure 4.1, add a product name or picture in the middle.
2. Let students hypothesize what constraints and criteria most likely frame that design space.
3. If there is time, have them do the same for another object.

Reflection Questions

You can debrief students with the following questions.

- What would happen if you took a specific constraint away?
- What would happen if you substituted a different criterion? How would the design possibly change?

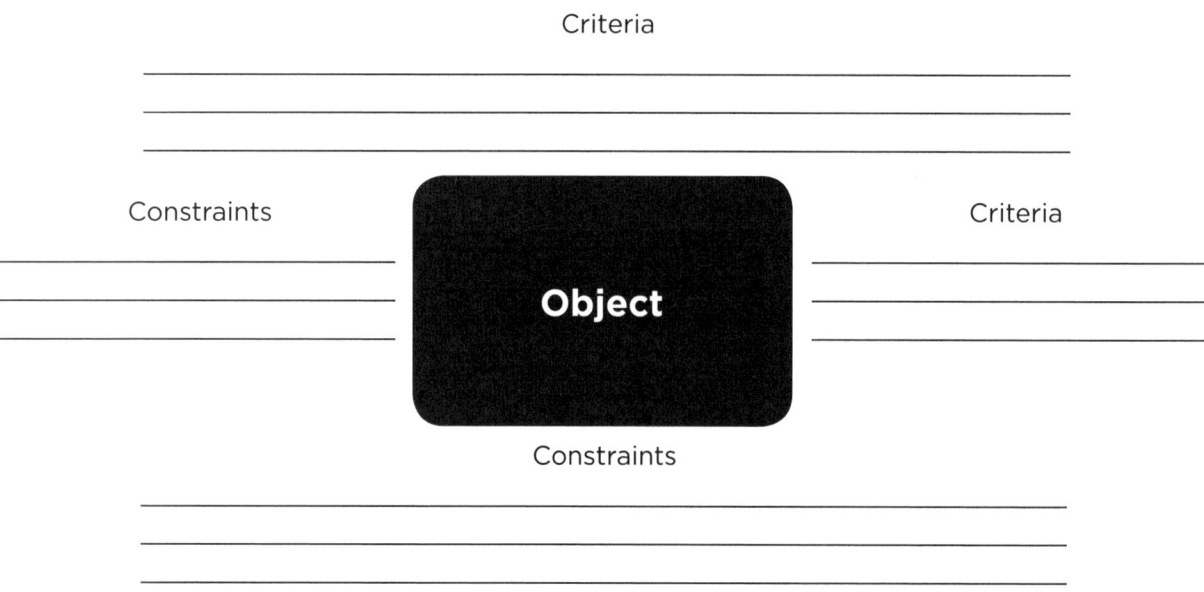

Figure 4.1: Reverse design space form.

*Visit **go.SolutionTree.com/21stcenturyskills** for a free reproducible version of this figure.*

A great follow-up discussion to this activity for identifying constraints involves somewhat of a paradigm shift. It basically shifts the scale of a constraint to highlight what could be done differently. For instance, if I said I was going to give you $10 and you had to spend it on one thing at once, what you would buy would probably be different than if I told you I was giving you $100,000 to spend at once. I am thinking something like pizza versus Porsche! The $10 constraint really limits your options.

Get creative. Make the constraint scale go from small to large, and keep it in the realm of things familiar to students. You can even go bigger and point out the rules of a game or sport are constraints, and have students explore what might change if a rule was removed. They will soon get the idea—constraints are very much limitations that help define possible solutions.

Name Your Pain Activity

Engineers and product developers will tell you that if you can address someone's pain point—their problem—your solution stands a good chance of success. Very few successful products were created in a vacuum just because someone thought they were good ideas. In most cases, needs come first, not designed products. Most successful designs focus on dealing with an issue facing the end user. For example, you probably would not have picked this book up if you didn't think something in it could help make your classroom learning experience different. And if this book assumed you have the time or inclination to throw out everything you do and replace it with a totally new approach, you would have stopped at the introduction.

> If this book assumed you have the time or inclination to throw out everything you do and replace it with a totally new approach, you would have stopped at the introduction.

The idea of trying small steps first and thinking of engineering design as a methodology to support skills-based, active learning in your classroom addresses the pain points of the many teachers I have worked with. They all cite limited time, a packed curriculum, and the pressures of trying to keep up with all that is new as significant pain points.

The simple Name Your Pain activity teaches a few things in a fun and engaging way. The first is the idea of empathy. Making Caring Common project members Stephanie Jones, Rick Weissbourd, Suzanne Bouffard, Jennifer Kahn, and Trisha Ross Anderson (2018) point out that:

> Children and teenagers naturally have the capacity for empathy, but that doesn't mean they develop it on their own. They learn how to notice, listen, and care by watching and listening to adults and peers, and they take cues from these people about why empathy is important.

Letting students share their pain points and giving them time to brainstorm possible solutions empowers students to value and care about someone else's challenges. A brief from the Harvard Graduate School of Education (Jones, Weissbourd, Bouffard, Kahn, & Ross, 2018) says *empathy* supports several positive outcomes we associate with positive learning experiences and environments. These outcomes include more classroom engagement, higher academic achievement, better communication skills, less aggressive behavior and bullying, and better relationships (Jones et al., 2018).

This activity also teaches students active listening—a good interviewing skill. Each student will get a chance to share his or her pain point with a partner. You must remind the student that when playing interviewer, he or she should listen to the interviewee's responses. It is basic human nature to jump in and say something along the lines of "I know. When that happens to me I. . . ." Think of your last cocktail party or faculty lounge conversation. Did anyone interrupt you, or vice versa? Remind students this is an interview, not a conversation. Developer Mark Suster (2013) says that the number-one rule in learning about your customers is that you can't be a crocodile, with small ears and a big mouth. Think big ears, small mouth for this activity. Students will get all the information they need by listening to their partners.

This activity will have students rapidly developing some possible solutions. You are bound to get some innovative (and sometimes even patentable!) solutions and have lots of fun. Everyone gets a chance to vent, and it provides a quick opportunity for some community building. Name Your Pain also provides a springboard for stressing the need to keep the end user in mind throughout projects.

The form in figure 4.2 is for middle and high school students. Elementary teachers can guide student pairs through it. In grades K–2, it might be helpful to do a similar activity as a whole class; let students interview you about a problem you face. Depending on their ages, they can then develop some ideas for a solution as a group or in smaller teams. If you teach middle or high schoolers, you may find it helpful to go through a sample activity with the entire class, citing one thing that bothers you every day. This helps them think about and see what might be useful questions. No matter what grade level you teach, allow some time for class sharing of innovative solutions when students have finished sharing with their partners.

This activity does not have to be tied to any specific project and can serve as a good team-building exercise. You can also modify it to focus on a specific problem for any engineering design challenge. Visit **go.SolutionTree.com/21stcenturyskills** to access free instructions to this activity.

Challenge

Interview another person to determine what is a problem for them or what bugs them.

Group Size

Two students

Time

Twenty minutes

Directions

These are the directions.

1. Give students some examples of pain points (things that bug them all the time) and discuss the idea that interviewing means asking questions, listening, and recording answers. The handout has some prompts but let them know they can use questions that they think are helpful.

2. Set a time for the initial interviews. Three or four minutes usually works for one interview, but adjust the time if the interviewees are still actively sharing. They get another three or four minutes when they switch roles.

Part 1: Interview your partner by identifying their pain point and asking them the following questions. Record their responses in the corresponding boxes.	
Pain Point:	
1. How often does this bother you?	2. Why is this a problem?
3. Have you tired anything to solve or fix this? What happened?	4. Is there anything else I should know about your pain point?
Part 2: Now switch roles and share your pain point with your partner. Working quietly by yourself, complete the following table to think about what a good solution should do. Once you are done, ask your partner for input or a possible ranking of what matters most.	
What I think, based on interview responses (in no particular order)	**What my partner thinks**
Part 3: Choose at least two ideas from the preceding list and try to create a solution that deals with both (or more). Working by yourself, draw a sketch or write a description. Share your solutions with your partner when you are both done.	
Sketch or describe possible design ideas:	

Figure 4.2: Name Your Pain activity form.

*Visit **go.SolutionTree.com/21stcenturyskills** for a free reproducible version of this figure.*

3. This activity is always best done in pairs. If you have an odd number of students, join in. This gives students a chance to learn a bit more about you and vice versa.

4. Working quietly in part two means no talking. Allow two or three minutes for this part. Then they can share their ideas with their partners.

5. In part three, allow three or four minutes for students to sketch or make notes about the solution to their client's pain point. Then have them share.

6. Ask volunteers to share the solution they created or the one that was created for them. Have them give the pain point first and then highlight how the solution addresses it.

Reflection Questions

You can debrief students with the following questions.

- How was the interview experience, and how did it influence your design?
- How did it feel to have something designed especially for you?
- What helped you to explain your solution?

Criteria Ranking Activity

It is important to have groups agree on the relative importance of the proposed criteria. I often accomplish this with a simple ranking process based on individual votes. You can use a matrix-based weighted ranking process for high school students if a complex design might be involved. This is rarely necessary, but it is a great connection to the real world of design and product development. TeachEngineering (https://bit.ly/2ITDc2K) has a great example of a complete matrix.

A criteria-ranking process lets students see the relative weight of various criteria due to vote tallies. Using a simple common object (separate from any large-scale design challenge) can help students see the value of developing criteria since the criteria often lead to a solution unique to their group. The following design scenarios work well for this short activity.

- Design a new student desk.
- Design a new classroom.
- Design new socks.
- Design a better backpack.
- Design new carry-on luggage.

Simply have the group come up with four to six criteria they think might be important. Have members make a list of criteria *in no particular order*. Then, give each student member a criteria-ranking activity form (you can call these *voting slips*) with two columns, one for rank and one for criteria (see figure 4.3). You can provide blank templates or have students create voting slips quickly on paper or index cards.

Rank	Criteria

Figure 4.3: Criteria Ranking activity form.

*Visit **go.SolutionTree.com/21stcenturyskills** for a free reproducible version of this figure.*

Students vote quietly, without any discussion. This restores some democracy and minimizes *groupthink* or the tendency to simply follow the loudest voice. Once they are done voting, students need to tally the votes to get one ranking. Figure 4.4 shows examples of students' voting slips as well as their final tally. It seems easiest to let students rank the criteria they think is the most important (number 1) and so on down the list. This means when votes are tallied *the criteria with the lowest score is the most important* (or number 1) criteria.

Figure 4.4 features Criteria Ranking activity forms and tally for the challenge of designing new carry-on luggage. Apparently, this group is going for stylish, lightweight luggage that may not last too long. Their solution would probably look different from a bag that is very durable and capable of different purposes. With practice, most students will get the idea that although the constraints may be the same for all groups (or designers), criteria are often different. This is what leads to the wide range of items around us.

Rank	Criteria
1	Appearance
2	Customizable
3	Lightweight
4	Has Accessories
5	Durability
6	Versatility
7	Safety

Rank	Criteria
1	Lightweight
2	Safety
3	Appearance
4	Durability
5	Has Accessories
6	Versatility
7	Customizable

Rank	Criteria
1	Has Accessories
2	Customizable
3	Appearance
4	Lightweight
5	Versatility
6	Safety
7	Durability

Rank	Criteria	Total
1	Appearance	7
2	Lightweight	8
3	Has Accessories	10
4	Customizable	11
5	Safety	15
6	Durability	16
7	Versatility	17

Figure 4.4: Criteria Ranking activity forms and tally.

*Visit **go.SolutionTree.com/21stcenturyskills** for a free reproducible version of this figure.*

Know Your Options

If you remember from chapter 2, the EDP moves back and forth between convergent and divergent thinking. Most of the teachers I have worked with find this phase, which is based on divergent thinking, challenging to successfully incorporate into their classroom. That is not surprising, since much instruction involves the teacher delivering a lot of information in a sequential format with a final, highly scripted assessment at the end of each unit. The traditional instructional format is a very linear and highly convergent model.

Many students will be uncomfortable in the wide open spaces needed for divergent thinking, and many will want to get directly to building, which is more concrete and defined to them. Show students a picture of Earth from space and remind them that we would never have the big picture if we never ventured into the unknown. Point out how crazy the idea of a round earth must have seemed centuries ago to someone who lived on the very flat plains and never traveled very far. Many great discoveries start out as "crazy" ideas.

Creativity can be tough to foster, but psychologist Ron Friedman (2015) says:

> In order to be creative, sometimes you need to consider some ideas that don't necessarily feel like they're on track with what you're trying to achieve. And so having all these ideas come into your mind because you're not quite as good at putting them off when you're tired can actually make you more creative.

Introduce some creativity and divergent-thinking exercises when students are tired (Friedman, 2015). These are great end-of-the-day activities.

Many teachers find getting their students to brainstorm is a challenge. Like anything, practice and some guidance from you can be really helpful. Samantha Scutieri, high school mathematics and engineering teacher, Union Catholic High School, New Jersey, says this:

> The brainstorming can be one of the hardest parts for them, because they have trouble getting away from the familiar. With creative prompts and activities, they start to think in a more ingenious way. (personal communication, August 8, 2018)

How do you get students to explore the crazy world of random ideas? Cindi May (2012), a psychology professor, points out when trying to think in a different way, distractions may be a good thing:

> This is where susceptibility to "distraction" can be of benefit. At off-peak times we are less focused and may consider a broader range of information. This wider scope gives us access to more alternatives and diverse interpretations, thus fostering innovation and insight.

Just think of those Friday afternoons when too many early mornings, a food coma, and weekend distractions become members of your classroom. Sometime after lunch might be perfect for creativity and divergent-thinking exercises. Try them first with fun, nontechnical ideas; these will more readily be part of your toolbox when you are ready for more extensive engineering design projects.

The following activities are great ways to break from the routine. They provide ways to get your students thinking beyond the boundaries we normally give them. They can work in any class or with any topic that involves coming up with new ideas, such as an art project, a new ending for a novel, different outcomes in historical contexts, or even generating scientific hypotheses.

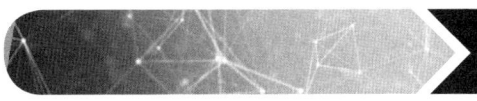

Stars and Stripes (and Dots) Activity

This is a fun, low-tech activity to get your students to embrace *synthesis* as a way to come up with new ideas. It is quick and occasionally results in patentable ideas! I like to use scrapbooking paper, but any colored paper or note cards will work. You can also have students add two (or three) columns of words on chart paper or a whiteboard and then let them choose one from each column to combine. It's quick, so it acts as a good warm-up activity before any creative endeavor.

Challenge

Combine the words you have chosen to create new things. You may combine the actual objects or features (for example, a rose has petals) or attributes (for example, a rose smells good) of the words.

Group Size

Two or three students

Time

Ten to fifteen minutes

Materials

These are the materials.

- Two or three kinds of paper with any sort of pattern on one side (stars, stripes, or dots, for example)

Directions

These are the directions.

1. The teacher, using two or three kinds of paper with a pattern (stars, stripes, dots) on one side, prints random words in columns on the white side.
2. Cut slips so that there is one word on each blank side.
3. Spread out the slips, pattern sides up, and have one student from each group pick one of each pattern. No peeking yet!

4. Once all student groups have each set of the slips, they turn the slips over and come up with as many ideas as possible that combine the objects or their attributes. They get three to five minutes for this. Then each group writes down or sketches ideas. Remind the groups that in all brainstorming activities *quantity* is the goal, not quality.

5. If they need help to get started, remind the groups that lots of things we use are combinations. The classic example is combining a phone with a computer to get a smartphone; combining a suitcase and a dolly (trolley) led to luggage with wheels; and combining a coin punch and a wine press led to the printing press. Your students can probably come up with more. Combining attributes can lead to even more options. For instance, socks and leaves could combine to create socks with removable leaf-like layers. Try not to suggest too many things.

6. Allow teams to share what they consider their coolest idea. I guarantee some will impress you!

Reflection Questions

You can debrief students with questions like the following.

- Was it hard to get started? If so, why do you think that was the case?
- Can you find some examples of things you use that combine features or attributes of two or more things?
- What was the most interesting idea you heard?

Chindogu Activity

Chindogu is a Japanese phenomenon. It is referred to as the art of *un-useless invention*, meaning these weird inventions are not quite useless, but not necessarily useful (Rogers, n.d.). Chindogu creations include items like chopsticks with an attached cooling fan for eating noodles, little umbrellas that attach to your shoes to keep them dry, and dusters that go on your cat's feet to clean your floors. Sometimes chindogu are not so useless or simply ahead of their time; the selfie stick was first developed as a chindogu exercise (Chindogu, n.d.).

> Chindogu creations include items like chopsticks with an attached cooling fan for eating noodles, little umbrellas that attach to your shoes to keep them dry, and dusters that go on your cat's feet to clean your floors.

Students love learning about and looking at chindogu, which highlights creativity without a purpose. Inventor Simone Giertz (2018), a self-professed overachiever, puts it beautifully in her TED Talk: "The true beauty of making useless things [is] this acknowledgment that you don't always know what the best answer is. It turns off that voice in your head that tells you that you know exactly how the world works."

The Chindogu activity focuses on both creative thinking and the ability to explain ideas with a physical prototype. Here is an activity that gives your students a break from the one-right-answer thinking so prevalent in education. It works for all grade levels after you introduce some examples. You can easily combine this hands-on activity with the Name Your Pain activity (page 105) or just have groups do it as a stand-alone activity.

Challenge

Given the materials available, create a chindogu.

Group Size

Two or three students

Time

Twenty to thirty minutes

Materials

These are the materials.

- Paper scraps
- Fabric scraps
- Paper clips
- Pipe cleaners
- Duct tape
- Any available odds and ends such as packing materials, different fasteners, and the like

Directions

These are the directions.

1. Student teams create their own chindogu.
2. Assemble a gallery of all of these marvelous inventions.
3. Students try to identify a possible purpose for the chindogu they did not create. They can write down any purposes they identify on sticky notes placed near the chindogu.

Alternatively, each group can share its creation in a short two-minute pitch to the class.

Try a modification challenge. Have groups of students select their favorite chindogu invention. It is great if they can print a picture of it. The challenge is to shift the invention into being a bit more useful. Give the groups time to brainstorm ideas and then sketch possible modifications to help un-chindogu the chindogu.

Reflection Questions

You can debrief students with questions like the following.

- Is it easier to create something when you know it does not have to have a serious purpose? Why or why not?
- Who is most likely to think your chindogu is helpful? (What is the possible target audience?)
- Can your chindogu be modified for different users or purposes? (The ability to be used by more than one type of person or for more than one purpose is often a selling point.)

SCAMPER This! Activity

The goal to brainstorming is to generate lots of ideas; it is very much about quantity over quality. But all brainstorms slow down at times. There are a lot of tricks to help keep ideas flowing. Stress that certain words, such as *if*, *but*, and *can't* are highly convergent and limiting. Listen to the groups in your classroom and help them to change direction, using words like *and* and *or*.

If groups need more prompts, SCAMPER (Eberle, 1996) is a time-tested acronym that provides ways to form lots of prompts. *SCAMPER* stands for substitute, combine, adapt, modify (or magnify or minify), put to other uses, eliminate, reverse (or rearrange). In his book, *Thinkertoys*, creativity expert Michael Michalko (2006) points out that you can manipulate all ideas into new ideas and designs. SCAMPER effectively provides multiple techniques for doing that.

One way to practice using SCAMPER is to pick an object or process, particularly one your students often complain about. Things that often fall into this category are student desks, cafeteria lines, and homework. I have even heard one student mention his baby brother, but we decided not to reinvent him! Figure 4.5 shows a sample from one of my workshops to help you get started.

Challenge: Redesign the plastic six-pack holder to be less harmful to the environment and marine life.		
S	**S**ubstitute	• Cardboard instead of plastic • Reusable holder • Edible holder
C	**C**ombine	Holder is built into reusable grocery bags
A	**A**dapt	Assume you can't eliminate anything; create a recycling campaign like the Box Tops for Education campaign—schools get money for every holder turned in.
M	**M**odify, magnify, or minify	• Have a break-away point in each ring • Make from materials so the elastic holders contract into almost a solid piece versus a ring when a can is removed
P	**P**ut to other uses	• Becomes a bracelet • Becomes a hair elastic
E	**E**liminate	• Small cans—only use large individual bottles • Design cans to clip together—no holder needed
R	**R**everse or rearrange	• One-piece biodegradable sleeve allows cans to slip in, creating a vertical arrangement of cans

Source: Adapted from Eberle, 1996.

Figure 4.5: SCAMPER This! activity example.

*Visit **go.SolutionTree.com/21stcenturyskills** for a free reproducible version of this figure.*

When the class shares, you are certain to hear a wide range of ideas—the whole point of the activity. This activity works best for grades 3–12; with some simplification and examples, it can work for grades 1 and 2. Visit **go.SolutionTree.com/21stcenturyskills** to access free instructions to this activity.

Challenge

Brainstorm solutions using SCAMPER prompts.

Group Size

Two to four students

Time

Fifteen to twenty minutes total (five minutes brainstorming and between ten and fifteen minutes for group sharing)

Materials

These are the materials.

- SCAMPER This! form (see figure 4.5)

Directions

These are the directions.

1. Each group uses an idea from the SCAMPER acronym (so every letter is used within the class), or the teacher can assign different letters of the acronym to different groups.

2. Pick an object or process, particularly one your students often complain about (for example, student desks, cafeteria lines, and homework).

3. Groups have five minutes to generate ideas for modifying or redesigning the object or process. Their ideas should be guided by the letter from SCAMPER that they were given or that they chose. If idea generation slows down, allow groups to pick one more letter to see if they can get things moving again.

4. Allow two or three minutes for each group to share its ideas with the class.

Reflection Questions

You can debrief students with questions like the following.

- Was there one letter in the acronym that was particularly helpful in terms of the number of ideas you could come up with?
- Did any of your ideas connect to two or more letters in the SCAMPER acronym?

Brainstorming During Projects

Brainstorming sessions within your projects may need some management skills on your part. Most teachers who try some version of the following find that students generate more ideas and that the process becomes more inclusive while minimizing groupthink. Always start any type of brainstorming by reminding your students of four things: (1) there are no bad or crazy ideas, (2) respect is rule number one, (3) this is really about quantity, not quality, and (4) no *ifs* or *buts* are allowed!

> Brainstorming is really about quantity not quality.

I have seen these three steps work well at all levels starting at grade 3.

1. Each student in a group gets between ten and twenty sticky notes. Each student gets all the same color.

2. Students get two or three minutes for students to generate ideas by themselves. No talking allowed!

   Explain that they can represent an idea by a note, phrase, word, or sketch on a sticky note. The goal is to fill up as many sticky notes as possible—one idea per note. Encourage this by pointing out highly productive students and groups. Don't be afraid to throw in a paradigm shift or prompt if things slow down. Paradigm shifts include quick suggestions such as *money is no object, add another function, it has to work for a giant*, or *it needs to make music*. Use any phrase or word that really shakes things up.

3. Now they can talk! Students go to their groups and arrange their notes in the center. They can look for patterns, similar ideas, things they all like, and other things.

   Usually some amazing plans start to come together and they see firsthand (through the sticky note colors) that everyone has ideas to contribute. This stage generally takes between ten and twenty minutes depending on the number of sticky notes they need to organize and how detailed the discussion is.

Develop a Solution

This stage of an engineering design project is the hands-on building stage. It is difficult to capture all of this stage in a short introductory activity, particularly when there is no specific design challenge or specified solution. It is useful, however, to focus on the value of developing a better understanding of several key ideas through a range of low-risk and time-efficient activities and exercises. This section highlights ways to focus on the following.

- What it means to make a prototype
- Modification based on test results
- Communication and collaboration skills

Prototype Time

Think of *prototypes* as formative assessment. Just as you use formative assessments to modify and adjust learning experiences, you should use prototypes to modify the planned final product. Prototypes should be indicators of what works, what needs improvement, and how close students' solutions are coming to addressing needs and meeting constraints and criteria. Most importantly, prototypes allow designers to check for user interaction as well as function.

> Think of prototypes as formative assessment.

Quick builds provide great experience with prototyping. These are quick ways to create a model of an idea and, in many cases, obtain some testing information to potentially make modifications. Quick builds model *rapid prototyping*, a product-design practice that allows for fast failure and redesign. In projects, I use quick builds as engagement experiences or hooks. Chapters 5 (page 125) and 6 (page 149) describe some quick builds with specific projects.

A great way to introduce the value of a prototype is to use an activity called Ready, Set, Design (https://bit.ly/2Fnd79f) from the Cooper Hewitt Design Museum of the Smithsonian Institute. This fifteen- to thirty-minute activity uses simple materials to introduce students to the value of a prototype to explain a solution (Shelly, 2011).

Encourage your students to use their imaginations and stress they should link some material properties between what they use and materials used in the actual solution, like those in the following list.

- Popsicle sticks can be lumber.
- Chopsticks can be dowels.
- Straws can be bamboo.
- Aluminum foil can stand in for most metals.
- Foam core sheets can be wood boards and drywall sheets.

Another great way to practice prototyping outside of an engineering design project is to substitute *prototyping* for any verbal way students describe a solution or outcome. For instance, if you sometimes ask students for a few paragraphs describing another outcome to a story or historical event, set up a *prototyping station* with simple materials and ask partners or teams to create a model with that new outcome. Asking students in kindergarten how they could help Cinderella keep her shoe on or how they might make a just-right tester for Goldilocks, engages them in a fun hands-on engineering design activity.

It is worth noting that a sketch can also be a prototype. If building an actual prototype is too complex, having the group create a large sketch on chart paper can be enough for members to explain or pitch their solutions. The sketches should be clear and easy to read from a few feet away. The sketches should include callouts for key features and students should also highlight how the design meets their criteria. Some of the projects in chapters 5 and 6 have sketches as the final project deliverables.

> If building an actual prototype is too complex, having the group create a large sketch on chart paper can be enough for members to explain or pitch their solutions.

Students can also create a public service announcement (PSA) or marketing video as their prototype. They can describe and illustrate processes this way. Just remember, the main reasons for prototypes are to act as visual aids for explanation and as a way to provide some testing or feedback data to make improvements.

The following Sketchy Ideas and Cardboard Carnival prototyping activities take about one class period. Once you try them, you will feel more comfortable trying larger-scale projects.

Sketchy Ideas Activity

If you can answer *yes* to the question, "Can I use this to explain my idea better?" you probably have some sort of prototype. If an object or a drawing also gives you some information about how well a solution works (either in terms of function or user interaction), you definitely have a prototype. Sketches are sometimes the simplest type of prototype to tackle first, and they are suitable for students in any grade. All you need is chart paper or some whiteboard space and markers or pencils. Sticky notes are helpful for letting students highlight key features.

Take one of the simple redesigns mentioned earlier (desks, backpack, or the like) and extend the activity to allow students to sketch their prototype. Have groups make the sketch big, so the class can easily see it when groups share. Allowing time for some feedback will give groups good information about their ideas. You can even try some of the gallery walk and pitch feedback ideas from chapter 2 (page 27) with this activity. Going low-tech as well as quick for a first prototype is a form of rapid prototyping. It allows for a nimble response and modification, and it's easy, cheap, and effective.

Cardboard Carnival Activity

A number of the schools I work with like the idea of the Cardboard Challenge (Imagination, n.d.) based on the Caine's Arcade video (http://cainesarcade.com). Of course, I like to put a little engineering design into the challenge. When you watch the video about a nine-year-old boy who creates an arcade out of discarded cardboard boxes at his dad's auto parts shop, you see that Caine is indeed an aspiring engineer. In fact, at fifteen years old, he planned on becoming an engineer. At one point in the video, Caine tests his prototype of a soccer (foosball) game and finds out customers think it is too easy. He beefs up the teams (by adding more plastic army men) in response to the feedback. Caine is modifying to optimize.

Rather than going directly to building large-scale arcade games, have your students create a mini version using limited materials. Then, allow time for feedback before they create the final version. Other students can identify what they like and don't like. They may even be able to try to play the game. Your class can even lay out a model of a full arcade with their prototypes. Most importantly, they will get an idea of whether anyone would come to their arcade or carnival before they go too far.

This activity works for students in grades 2–8.

Challenge

Build a mini-carnival game with some simple materials and the most important tool you have: your imagination!

Group Size

Two to four students

Time

Between forty and sixty minutes

Materials

These are the materials for each group. Consider adding materials (more of those that are listed here) for elementary students.

- Cardboard Carnival student handout (at **go.SolutionTree.com/21stcenturyskills**)
- Two pieces of 10 × 10-inch cardboard; substitute foam core or heavy card stock if preferred
- Six inches of duct tape (variety of colors and designs)

- Markers (selection of additional colors available for last-minute design details)
- Two pairs of scissors
- One ruler (to help with planning and cutting; can be omitted)
- Five pieces chosen from this list: an assortment of paper clips, binder clips, poster pins, pipe cleaners, straws, toothpicks, and other small objects

Time

Twenty minutes for building

Directions

These are the directions.

1. Teams do some planning, considering what games they like and what they think they can construct with the given materials.
2. When the teacher says "Time," two people from each group go get the materials.
3. Students have fifteen or twenty minutes to create their Cardboard Carnival mini-game.
4. Each team appoints a pitch person who can get everyone excited about the game in one or two minutes. That person shares with the class.
5. After every team shares, take a few minutes to think of items or changes (modifications) that would make a larger-scale model better. Fill in the back of this paper with a sketch and your modifications.

Reflection Questions

You can debrief students with the following questions.

- What modifications or additional materials would be helpful?
- What did you learn about your design as you built this prototype? What did you learn when you shared it?

The Cardboard Carnival activity modifications form in figure 4.6 can help you to get started.

Modification	Reason for Modification

Figure 4.6: Cardboard Carnival activity modifications form.

Visit **go.SolutionTree.com/21stcenturyskills** *for a free reproducible version of this figure.*

It's important to keep in mind the true value in prototyping quickly or letting students explore ideas with quick builds: *providing a low-risk opportunity for failure.* A time restriction must be in place to make it unlikely students will get things just right. Whether something fails due to physical testing (such as collapsing or sinking) or because it does not engage the intended user as expected,

A time restriction must be in place to make it unlikely students will get things just right.

there is a lot to learn. This is *failing forward* (Maxwell, 2007). Encourage your students to fail, but please don't end it there. Allow time for either planning modifications or for actually making modifications. That is how you give failure value. This kind of learning goes a long way in supporting resilience and lifelong learning.

Case studies and failure analyses are conducive to problem-based learning. Students can do much of the background research and preparation outside of class. However, excellent opportunities for learning technical concepts and highlighting ethics and values issues are inherent in classroom debates and discussions. You can tailor case studies and failure analyses for specific grade levels. Both are excellent methods for introducing the need for ethical decision making, social responsibility, and systems thinking. Depending on the grade level and class focus, you can also cover an appropriate amount of technical material. Visit **go.SolutionTree.com/21stcenturyskills** for case studies about the *Challenger* and the *S.S. Eastland* disasters to share with students.

Modification Time

Part of learning from failure is thinking about how to modify something to make it better. Practicing modification might be one of the most important things you can do in terms of life skills (and in terms of really understanding how our designed world comes into being). In addition, students learn to appreciate the fact that their solutions do not need to be perfect in the beginning, and will begin to value the iterative process. Doing simple activities that provide some opportunity for modification is valuable because it makes modification a natural part of longer activities. You can add a quick second round for modification to any of the building activities given earlier in this chapter if you have time, or you can try my favorite activity specifically for modification, ModiFly.

ModiFly Activity

One of the issues in making modifications is student prototypes often get pretty beaten up during physical testing. Boats sink, buildings collapse. The ModiFly activity uses an item students can easily make multiples copies of before testing, so they have enough subjects for modification. Low-tech, low-cost paper airplanes are perfect and also allow you to make connections to the curriculum. For example, it can be a social studies project related to the highly iterative EDP the Wright Brothers followed. It can be a physical science or physics activity dealing with the forces of flight and aerodynamics. Or it can be a simple engineering activity designed to highlight the value of testing and modification.

Because of the successive modifications, the activity as described works best for students in grades 5–12. Younger students can follow a more teacher-led approach with fewer modifications if desired.

Challenge

Make a simple paper airplane that can fly long distances.

Group Size

Two or three students

Time

Allow one or two classes periods (depending on how much modification you allow).

Materials

These are the materials.

- Six sheets of copy paper
- Transparent adhesive tape
- Paper clips
- Small sticky notes

Directions

These are the directions, including students' test parameters.

1. Each group makes six paper airplanes. They should be close to identical, very simple, and numbered (either on the wing or the body, also known as the fuselage). They put aside the fifth and sixth planes for the final stages of the activity, starting at step 8.

    I suggest allowing just one sheet of printer paper per airplane at this point. Point out to students that simple designs are always better because it is easier to identify what does and doesn't work. The Wrights first worked with simple gliders to understand and optimize different wing shapes before adding complex control systems and an engine (Smithsonian National Air and Space Museum, n.d.).

2. Have students create a data table for testing results.

3. Set up a testing area. Limit random flights. Allow the students to test the first model to see how far it goes. You may want to allow three tests to support more robust data. Students should record the distance to the first landing point.

4. Allow students to make *one* modification to plane number two. Each of the following counts as one modification.

    - One change in shape of the wing or fuselage (body of the plane)
    - One paper clip
    - One inch of tape
    - One sticky note

5. Students should record the modification on their data table and retest.

6. For round three, students should take another new plane (plane number three) and make one modification. Test and record.

7. The same should happen for round four, again with a new plane (plane number four). Test and record.

8. For round five, students choose plane two, three, or four to make one more additional modification to. Test and record. This step allows students to practice a higher

level of design by synthesizing the best features of multiple designs. From middle school, synthesis of both design ideas and features is highlighted in the engineering Disciplinary Core Ideas of the NGSS (NGSS Lead States, 2013).

9. For round six, students can choose to add to the plane they tested in round five or to one of the other planes.

10. Using one of the two untested planes (planes five and six from the original construction phase), students make the best plane possible using all the modifications tried, and then test it.

    You can allow one more modification to the last remaining plane. It is basically a backup, but if it they have not used it, they can at this point.

11. Allow some time for discussion of improvements, possible testing discrepancies (hard to throw it the same way, make the exact same folds, and so on) and anything students may have learned about the physics of flight and the value of controlled modification.

Reflection Questions

You can debrief students with the following questions.

- What were some of issues you faced in testing? Was it hard to throw it the same way, to make the exact same folds, or something similar?
- What did you notice about wing shape? About overall body shape?
- Did your modifications lead to a better plane? Are there any additional modifications that you think might have worked better?

Communication and Collaboration Time

It makes sense to focus on communication and collaboration together. Teamwork is hard if you do not know what others are thinking. Many of this chapter's earlier activities support collaboration, but it is sometimes helpful to include an activity with a strong focus on clear communication. Encouraging students to write pictorial instructions like they do during the No Words activity (page 146), or highlighting oral instructions, helps them recognize how often we are less than clear when explaining procedures and ideas to others.

The LEGO Person Activity

The following challenge is a great team-building activity; you can use it in English language arts classes that focus on communication strategies. This activity also has a significant spatial reasoning component. Chapter 6 (page 149) discusses the value of spatial reasoning activities. A lack of spatial thinking practice is increasingly linked to a range of students' academic struggles (Newcombe, 2013). The goal of the LEGO Person activity is for students to practice their verbal communication, as well as their visual-cue, spatial-orientation, and collaboration skills.

This works well for all ages, with younger students using DUPLO or other similar blocks.

Challenge

Construct a replica structure based only on someone else's verbal instructions.

Group Size

Four students

Time

Approximately thirty minutes

Materials

These are the materials.

- Seven sets of ten to fifteen identical LEGO bricks, DUPLO pieces, or similar building blocks (each set in its own bag)
- Seven small sticky notes
- Seven markers
- Stopwatch or phone for timing (teacher)

Directions

These are the directions.

1. Prior to class, put together the sets of blocks and include one sticky note and one marker in each bag. Take one bag and build a LEGO person. Give him a happy or sad face on a sticky note.

    To increase difficulty, leave out one or two LEGO pieces. Don't let students know which ones, but do tell them you left some out.

2. Prior to class, hide the LEGO person somewhere students *can't* see him from their tables, but where observers (away from their group tables) will have a good view.

3. When students are in their groups, distribute the bags.

4. Provide a copy of these instructions, available at **go.SolutionTree.com /21stcenturyskills**, for each group and go over them with students.

5. Each group appoints an observer.

6. When the teacher says "Go," all observers go to the LEGO person. They have one minute to observe him. No one can touch him; their hands must be behind their backs.

7. Observers return to their group when the teacher says "Stop." When they are with their groups, *the observers must stand with their hands behind their backs.* When the teacher says "Go," the observers have two minutes to tell the team how to build the LEGO person.

8. At the end of two minutes, each group chooses a new observer.

9. When the teacher says "Go," they will repeat steps 6 and 7. No building can happen while the second observer is gone.

10. When the second observers have provided instructions, the first observers can join in building.
11. Repeat two or three more times, with a new observer every time.
12. After four or five rounds, the teacher unveils the original LEGO person.

Reflection Questions

You can debrief students with the following questions.

- What was the hardest thing to describe verbally?
- What were some strategies your team used to work together?
- Is it easier to follow instructions that are written?

Doris Treuer, a seventh- and eighth-grade science teacher at the Academy of Our Lady, in New Jersey, says the following about projects requiring engineering design:

> Engineering design projects have given my students the confidence to try, fail, and try again. They are beginning to realize that failure is not the end to their work, it is a necessary part of the engineering design process. I often refer to an inspirational poster in my classroom which states, "You will never make a discovery if you are always afraid to make a mistake." (personal communication, September 30, 2018)

Going Forward

Don't be afraid to *do*—just don't do too much at once. Most of us walked before we ran, testing our balance and learning that we could get back up when we fell. This chapter is about walking before running into projects. Try using some of these activities before committing to full-scale engineering design projects. New ways of thinking and doing are challenging for all of us, and that first step can be a bit scary. Remember to make those first steps small, and commit to taking them. After all, this is something you and your students can only learn by doing, and small steps make falling hurt less!

CHAPTER 5
Introducing Projects for Elementary School

An ounce of action is worth a ton of theory.

—Friedrich Engels

Think of things you can easily remember—something made these things more relevant, useful, or alive. Whether it was learning how to ride a bike, using various computer applications, playing a game, or finding the best way to get to work, much of our lifelong learning involves *doing*. No one would dream of teaching young children to swim by setting them at desks and showing them notes and graphics of different strokes. You need the sensation of developing a way to breathe and move when surrounded by water to truly learn how to swim. You need to sputter and blow some bubbles to learn how to avoid inhaling water; you need to feel the resistance of the water as it pushes back to figure out how to use your arms and legs effectively. And, as parents and teachers, we would never let a novice swimmer in the water without being nearby for support and assistance.

If you hope to develop students who know how to swim in all sorts of situations, you need to provide an environment that gives them a chance to sputter as they sort out the best way forward. Encourage the active application of the theories and concepts you help them learn. Let students discover the value of a support network and feedback as they work to make sense of information. If you hope to teach creative problem solving, students need to "jump into the water" of a good challenge. An increasing number of recent studies highlight the value of active early STEM experiences. A Community for Advancing Discovery and Research in Education brief says that, among other things, STEM experiences naturally propel students toward literacy and language competencies (Sarama et al., 2018).

Chapter 3 (page 59) provides some guidelines and resources for structuring your own projects. This chapter and chapter 6 take you through the overall design and structure of some projects you can use in various disciplines at various grade levels. Each of the projects focuses on different curricular concepts and skills, but all share the EDP framework. Project plans explain the EDP steps. In some instances, more than one project focuses on the same topic; in those cases, a project summary provides an overview of both projects. In addition, you can use some ideas for longer activities that support engineering design thinking and practices if you are still not ready to jump into a full-blown project.

Best Practices for Elementary Students

Although students in grades K–2 are often our most imaginative, enthusiastic, and natural engineers, they can be the most challenging to manage and assess. K–2 students may not function well in teams, and they can't document a process in full. They will benefit from a simple way of documenting steps and thinking about planning and testing in addition to building. Following a process and learning to collaborate are often new experiences for these grades K–2 engineers. It helps to script your design challenges with key steps that help students do three things: (1) focus on planning, (2) stop to think about *why* before *doing*, and (3) understand feedback and failure help them to make things better.

> If you hope to develop students who know how to swim in all sorts of situations, you need to provide an environment that gives them a chance to sputter as they sort out the best way forward.

In many cases, I find drawing analogies to what students are familiar with can help them be innovative with intention. This works at all levels, but seems particularly helpful in grades K–5. For example, clothes serve some of the same purposes as the fur on some animals; and sorting and organizing books and toys are similar to the way the periodic table organizes elements and the classification systems categorize plants and animals. Think of more ways that engineering inspiration comes from nature and science. Planning *something* so it is like *something else* is much easier than developing a plan from a blank slate.

Students in grades K–2 often work better side by side than in groups. However, I have seen some kindergarten students work well in pairs if it is the norm for a range of learning experiences in their classroom. Many PBL practitioners believe group size should equal grade level for young learners (Lachapelle & Cunningham, 2014). After grade 4, groups of four or five students are effective for most of the projects I have designed. To introduce the idea of a collaborative effort, consider giving different pairs of students different jobs related to the overall project. For instance, if students are working on a project to make sure the Prince can find Cinderella (see Engineering Happily Ever After, page 131), you can brainstorm multiple solutions as a class and then have different groups develop different types of solutions. One group can develop sticky shoes that don't fall off; another can develop some sort of light or sound to act as a beacon for the Prince to follow; and a third group can develop some way for Cinderella to leave a trail, like confetti falling from her dress. All sorts of solutions might work and by creating all the multiple options, your class will learn they can always engineer more than one solution.

Have your students make a simple sketch before any building starts. For kindergarten students, the drawing will probably not look anything at all like what they actually make, but expect they will get better with experience. Have students begin to add *callouts* or separate little sketches of the really cool things about their ideas. Students are more likely to try to model these ideas and believe their ideas are special. As students get into grades 1 and 2, add a focus on learning something from testing and feedback to their understanding of following a process. It will help if students value learning from mistakes. This step also begins to introduce the iterative nature of engineering design and how to try ideas as students develop solutions. Students in grade 2 are usually able to understand the idea of modifications and like having the opportunity to change and improve their designs. Just be careful to not let them start all over every time something doesn't work too well.

Assessment for elementary students typically focuses on skills and is generally based on individual performance, progress, and competencies. In my experience, assessment generally follows the existing approach for other learning activities in your school or district.

Some projects that work well for elementary students include those connected to literature, such as Huff 'n Puff (page 128), Just Right (page 132), or anything modeled on Engineering Happily Ever After (page 131). Disassembly- and assembly-type projects also work well at this level. Taking things apart, or *reverse engineering*, is a great way for elementary students to learn about the designed world. Always ensure there are no sharp component pieces or electronics, such as capacitors, to worry about. A simple assembly-line activity (for instance, a pretend toy factory) also helps students focus on planning and collaboration. You can scale down the How Does Your Production Line Rate? activity (page 175) for this age group.

Many K–2 teachers I have worked with try a pictorial instruction (No Words activity, page 146) project or activity. This type of activity reinforces spatial reasoning and helps students develop planning and collaboration skills. Be careful not to use too many pieces which will create too many steps for them to describe. Encourage your students to name their creations. Repeated opportunities to try projects like this one leads to activities that involve creating written process instructions in grades 3 and up.

Most teachers try projects based on some of the models and plans in chapters 5 and 6 first. But, don't be afraid to modify those or even create your own projects. Use the "Project Planning Template" reproducible (page 206) to create or modify projects, or visit **go.SolutionTree.com/21st centuryskills** for live links to online resources.

Many teachers start with one of the projects or activities in this chapter and chapter 6 and then move on to use the ideas from the earlier chapters to create both new and similar projects as they adapt learning experiences to their own needs and those of their students. The intent of both chapters is to make those first steps toward change a bit more manageable.

Overall Approach

Each of these projects evolved from the project-planning process described in chapter 3 (page 59). Many made their first appearance on a form similar to the "Project-Planning Template" reproducible (page 206) in appendix B. They all begin with an engineering design challenge, tie into key curricular concepts, and focus on specific skills and key EDP steps. In most cases, the project suggests background instruction, possible outcomes, resources, and assessment. Projects are grouped by the main discipline or subject they would most likely live in. Most of the projects in this chapter have several variations and you can easily modify them all to fit your needs. You can use different books for Engineering Happily Ever After (page 131), different packaging designs or requirements for Building a Better Box (page 135), and all sorts of inspiration from nature for biomimicry-based projects.

These project overviews are not full lesson plans; they are synopses and the overall guidelines are a road map to get you started moving forward. Fill in the details by planning the stops and side trips along the way to meet your students' needs or use the projects as templates or model itineraries to plan your own project journeys.

English Language Arts–Based Projects

Using a book or story as the springboard for an engineering design challenge creates an opportunity for your students to connect to what they have read. In working to solve characters' problems, students will look at the story from different angles to understand their end user, and by considering multiple options they will learn there are often choices in the story line.

The first project, Huff 'n Puff, is essentially a civil engineering project for grades K–5 students. Next is an engagement activity called Engineering Happily Ever After (page 131) that you can use to generate projects based on the book or story of your choice. Teachers developed the Just Right project (page 132) as the result of an exercise using Engineering Happily Ever After. In that project, Goldilocks faces an ergonomics challenge as she struggles to find different things that fit her taste and size criteria.

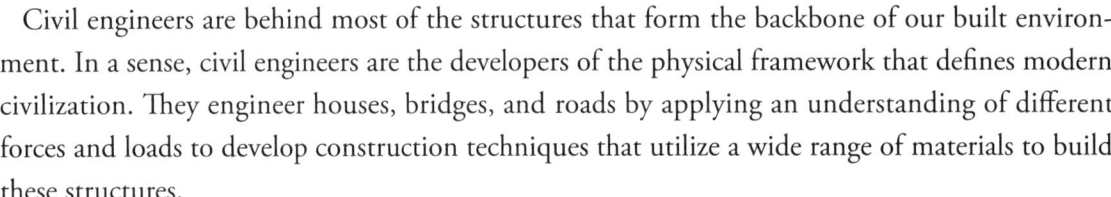

Huff 'n Puff: Disaster-Resilient Housing (Grades K–2 and Grades 3–5)

Civil engineers are behind most of the structures that form the backbone of our built environment. In a sense, civil engineers are the developers of the physical framework that defines modern civilization. They engineer houses, bridges, and roads by applying an understanding of different forces and loads to develop construction techniques that utilize a wide range of materials to build these structures.

Most consider civil engineering to be the oldest of all engineering professions, since it has the basic human need for shelter at its core. Basing a first engineering design project on civil engineering is often an entry point for teachers because the products of civil engineering are tangible, large-scale structures. Young engineers have lots of models to learn from in the world around them and projects provide the satisfaction of creating a model of a large-scale structure. If you need some inspiration for your students, the Institute of Civil Engineers has a great website (https://ice.org.uk) highlighting civil engineering projects around the world.

Huff 'n Puff challenges students to learn from the failures and successes of some famous civil engineers using *The Three Little Pigs* (American Literature, n.d.). Students in grades K–2 are asked to plan, build, and test structures using some of the shapes and letters they identify in structures around their school or in pictures. Planning is a challenge for this group, but it should become an expected part of the overall process. The idea of doing things differently after a failure is an important one to introduce at this point since it highlights the value of failure framed as a learning experience.

Students in grades 3–5 can take a deeper look at the impacts of storms and their effect on different types of housing by looking at structures from around the world made of straw, wood, and brick. They can identify the constraints on design due to available building materials, and they can look at the connection of architectural design to culture. Students can also explore the idea of designing housing that can be more readily rebuilt to make recovery faster and communities more resilient in the face of natural disasters. The use of local materials, reinforced cores, and innovative techniques support the design of structures quickly rebuilt after suffering less-costly damage. For grades 3–5 students, I always suggest increasing the emphasis on the idea that constraints and criteria frame the designs; they should be the primary focuses of engineering design. The other added skill to focus on is a more formal approach to learning from failure and making modifications to design. See figure 5.1.

> For grades 3–5 students, increase the emphasis on the idea that constraints and criteria frame the designs.

Project Title: Huff 'n Puff	**Topic:** Civil engineering; disaster-resilient housing	
Grade Level: K–2: Shapes and letters 3–5: Cultural connections; resilient design; forces and materials		**Estimated Class Time:** K–2: Three hours 3–5: Five hours
Challenge: Design and build a house that can either resist or be easily rebuilt after a wind storm. K–2: Use some building ideas from things around you (shapes and letters). Grades 3–5: Get inspiration from materials and housing designs around the world.		
Curricular Connections	**Skills Focus**	
ELA—inspiration and connection to children's literature **Mathematics**—2-D planning to 3-D construction; shape identification **Social Studies**—cultural meaning of home; knowledge of other places **Physical Science**—forces and material properties **Earth Science**—weather events; wind **Art**—architectural design	**Critical Thinking**—choosing materials; identifying key needs; resistance versus resilience **Creativity**—architectural design **Spatial Reasoning**—planning in 2-D and building in 3-D **Global View**—meaning of housing in cultures **Collaboration**—team or group work	
Overall Plan		
Hook	Book tied to project	
Quick Build	Hands-on engagement—Paper Tower of Power activity (page 96)	
Background Instruction	• Forces • Weather and wind • Optional—architectural drawing and plans	
Background Research	• Shapes and letters we see in buildings • What houses look like around the world • Different materials used in buildings • What home means	

Figure 5.1: Huff 'n Puff project plan.

continued ▶

Engineering Design Process	
Know Your Problem • Know your end user • Identify constraints • Define criteria	• Students discuss and research types of housing; why different people like different houses. • Identify materials used in the real world and materials available for a model; how much space you will have? • What makes a house a good home? • What is most important?
Know Your Options • Research housing and design • Brainstorm	• See Background Research. • Use a variety of techniques to get students to consider innovative designs.
Develop a Solution—Part One • Choose a design • Identify needed materials	• Students settle on best design; older students discuss how it meets constraints and criteria. • Stress range of choice and need for planning more with increasing student age.
Develop a Solution—Part Two • Create a plan (make a sketch) • Build the house	• Students develop a 2-D sketch of their proposed house. • Students construct a 3-D model; expect older students to follow the plan more closely.
Develop a Solution—Part Three • Test the house • Plan some modifications to make the house better	• Grades K–2—Use a hair dryer (disguised as the wolf) to try to blow the house down. • Grades 3–5—Use a hair dryer or fan at different speeds to simulate a windstorm. Investigate perpendicular, parallel, and edge-on direction of the wind if houses stand. • Older students can record results and conduct some analysis. • Modifications—If houses are in reasonable shape, students can make modifications directly; otherwise have them suggest planned modifications. Students in grade K–2 can do this as a group discussion; students in grades 3–5 can complete a more formal modification form. Expect increased connection of modifications to test results and key concepts as student age increases.
Develop a Solution—Part Four • Communicate your results	• Older students can make a brief presentation discussing why they built their house a certain way. It should connect to local norms, materials, and climate and other weather conditions. Expect students to begin connecting to criteria and constraints. Encourage students to discuss testing results and ways to make it better. • Students in grade K–2 should have a chance to share what is special about their house and what they think might make it better.
Group Size	
Grades K–1: Have students work in pairs; teams are not effective at this age. **Grades 2–3:** Introduce groups of three. In grade 3, begin introducing jobs such as project manager, civil engineer, and architect. **Grades 4–5:** Groups of four generally work. Students should begin functioning effectively in teams; jobs are a must.	

Assessment
Grades K–2: • Include an assessment component that reflects how well students make and follow a plan. • The Buck Institute of Education offers a free Creativity and Innovation rubric (https://bit.ly/2xnMNZ8) for this level. Consider using or adapting it. • Although having individual and group components to grade is unrealistic at this point, emphasize to students that working as a team is important. Formatively assessing and monitoring their teamwork skills are helpful. **Grades 3–5:** • Begin to use a portion of the grade for the project to reflect the need for collaboration and teamwork. Collaboration skills should become a more significant assessment component by grade 4 and as they do more projects. • Identify curricular content understandings to assess, along with process components. At this age, students should begin to show understanding of the elements of problem definition (end user, constraints, and criteria). There should be evidence of students considering various options. There should also be an increasing focus on ways to test, along with using testing information to make modifications, as students get to grade 5.

Engineering Happily Ever After (Grades K–5)

A good book almost always has a good problem. Engaging stories often center around challenges that characters must overcome. Many characters could use some help meeting those challenges and solving those problems. Giving students the opportunity to engineer solutions to help book characters brings students into the book and engages them as part of the story. If you do a good job of stressing the need to really understand the end user, "story engineers" have little choice but to delve deeper into the context and nuances of plot and character development. Interacting with a book is a far more active form of learning than summarizing the content or plot development.

Students in grades K–2 can start with fairy and folktales; students grades 3–5 can look at challenges characters face in countless young reader books. The first step is to allow some student choice in identifying the challenge. Think about it—who has the real problem in *Cinderella* (Grimm & Grimm, 1812/2013)? Depending on your perspective it could be Cinderella, who might need a watch, or it could be the Prince, who would clearly benefit from some sort of tracking device. It may even be the Fairy Godmother since she has quite a bit to do with limited resources and time. Allowing students to identify the end user is a first step toward their ownership of the project and part of the problem-definition process.

> Giving students the opportunity to engineer solutions to help book characters brings students into the book and engages them as part of the story.

You can begin to engage students with figure 5.2 (page 132). This worksheet is designed to help students further define the problem or challenge by identifying criteria and constraints prior to brainstorming multiple solutions. Grades K–2 students will probably do better in a large-group response format, while grades 3–5 students can use the worksheet. Depending on how much time you allot for the activity or project, students could build, test, and modify prototypes of solutions.

> **Name of the story:**
>
> **Identifying the end user:** Who is the character with the problem?
>
> **Defining the problem:** Describe the problem he or she is facing.
>
> **Identifying constraints:** What are some things that limit the character's choices? Try to think of at least three.
>   1.
>   2.
>   3.
>
> **Developing criteria:** What should a good solution do? How should it work? Try to think of at least three ideas.
>   1.
>   2.
>   3.
>
> **Brainstorm solutions:** What would you do if you could do anything? Write your ideas on sticky notes or make a drawing of them on a separate sheet of paper.

Figure 5.2: Engineering Happily Ever After worksheet.

*Visit **go.SolutionTree.com/21stcenturyskills** for a free reproducible version of this figure.*

Just Right (Grades K–5)

The Just Right project, developed from the Engineering Happily Ever After approach, resulted in a consideration of ergonomics as Goldilocks looks to find the bed that is just right. (You can see the project plan in figure 5.3, pages 133–135.) *Ergonomics engineers and experts* are the people who work to make the objects we use fit our bodies in terms of comfort, safety, and ease of use, among other factors. Have students create a pen or pencil far too large in diameter by wrapping it in bubble wrap or other materials. This will illustrate to them how we assume objects will fit us. Many students will connect this experience with steps that have too high of a rise, chairs that are too small, or tables that are too high or too low. Students could complete an activity to measure the height and depth of chairs, desks, and even stairs. This will also show them these dimensions are standardized.

You can also conduct a more formal data-based exploration if you want a strong mathematics component.

Project Title: Just Right	**Topic:** Design a stool to hold you by exploring ergonomic design; forces on structures	
Grade Level: K–5: Varying levels of prototyping, mathematics content, planning Note: Grades 3–5 students can create full-scale stools from cardboard; it will be a challenge for grades K–2 students to create stools that hold a teddy bear.		**Estimated Class Time:** K–2: Six classes 3–5: Ten classes
Challenge: Can you build a stool that will be "just right" for Goldilocks (life-size, grades 3–5) or for her teddy bear (grades K–2)?		
Curricular Connections		**Skills Focus**
English Language Arts—inspiration and connection to children's literature **Mathematics**—measurement; 2-D to 3-D planning **Social Studies**—evolution of design to fit human needs **Physical Science**—forces and materials **Art**—aesthetics of furniture design		**Critical Thinking**—choosing materials; identifying key needs; analyzing effective designs **Creativity**—aesthetics of design **Spatial Reasoning**—going from 2-D planning to 3-D models; human interaction with designed environments and objects **Collaboration**—working as a team; recognizing common human factors **Communication**—presenting the final structure; key features and ergonomic concerns
Overall Plan		
Hook	*Goldilocks and the Story of the Three Bears* (Southey, Opie, & Opie, 1980) or any other book where a device or implement needs to fit humans	
Quick Build	Twenty minutes—make the best possible three-legged stool from four sheets of printer paper and two large (4 x 6-inch) index cards; 6-inch minimum height; 6-inch minimum diameter or diagonal seat; test the stool to see what it can hold—stuffed animals, books, and so on. Students may use tape and scissors. (See the Three-Legged Stool activity, page 98, for more details.)	
Background Instruction	• Forces • Center of mass, equilibrium, and stability • Material properties • Ergonomics	
Background Research	• Different furniture designs • Furniture-building materials • Development of ergonomics; ergonomic standards	
Engineering Design Process		
Know Your Problem • Know your end user • Identify constraints • Define criteria	• Choose someone to design for (in this case, a child or teddy bear); investigate what the end user likes to sit on, and what sizes are comfortable. • Use the Engineering Happily Ever After worksheet (figure 5.1, pages 129–131) to identify product or project constraints. • What makes the stool one that you or Goldilocks would like?	

Figure 5.3: Just Right project plan.

continued ▶

Know Your Options • Research • Brainstorm	• Investigate furniture (chairs and stools) design; principles of ergonomics. • Use a variety of techniques to get students to consider innovative designs.
Develop a Solution—Part One • Choose a design • Identify needed materials	• Students settle on the best design; students in grades 3-5 discuss how it meets constraints and criteria. • Stress range of choice and need for planning more with students beginning in grade 2.
Develop a Solution—Part Two • Create a plan (make a sketch) • Build the stool	• After some rapid prototyping activities, develop a sketch of the planned stool. • Obtain materials and build the stool; keep material choice simple. (The teacher should suggest using cardboard, white glue, or tape and also provide some items that allow designers to decorate.)
Develop a Solution—Part Three • Test the stool • Plan some modifications to make it better	Depending on the age group and whether a full-size or small-scale stool is built, students should test for the following. • How does it look? • Is it stable and level? No tipping allowed. • Can it hold the planned weight? 　+ Start with smaller load. 　+ Work up to a full load—teddy bear, teddy bear with weights, student. • If testing a life-size stool, have two spotters present to support the student. • Is it comfortable? Students who make life-size stools should get feedback from a few testers concerning comfort. • Modifications 　+ Does the seat need to be made differently, more comfortable? 　+ Are the legs strong? Can they be reinforced? 　+ Is there one thing you can do to make your stool better?
Develop a Solution—Part Four • Communicate your results	• Sell your stool to Goldilocks! Have students develop an ad or marketing campaign to convince Goldilocks this stool meets all of her "just right" criteria.

Group Size

Grades K-1: Have students work in pairs; teams are not effective at this age.
Grades 2-3: Introduce groups of three or four. In grade 3, begin introducing jobs such as project manager, civil engineer, and architect.
Grades 4-5: Students should begin functioning effectively in teams; jobs are a must.

Assessment

Grades K-2:
- Include an assessment component to reflect how well students make and follow a plan.
- PBLWorks offers a free creativity and innovation rubric (https://bit.ly/2xnMNZ8) for this level. Consider using or adapting it.
- Although having individual and group components to grade is unrealistic at this point, emphasize to students that working as a team is important. Formatively assessing and monitoring their teamwork skills are helpful.

Grades 3–5:
• Begin to use a portion of the grade for the project to reflect the need for collaboration and teamwork. This should become a more significant assessment component as students do more projects. • Identify curricular content understandings to assess along with process components. At this age, students should begin to show understanding of elements of problem definition (end user, constraints, and criteria). There should be evidence of students considering various options. There should also be an increasing focus on ways to test, along with using testing information to make modifications by grade 5.
Resources
The Push and Pull activity (at **go.SolutionTree.com/21stcenturyskills**) is great for learning about materials.

Visit *go.SolutionTree.com/21stcenturyskills* for a free reproducible version of this figure.

Mathematics-Based Projects

Mathematics can become a key part of many engineering design projects. Adding a cost component, scaled drawings, and 2-D to 3-D planning can be part of a wide range of projects, including the Huff 'n Puff and Just Right projects in figures 5.1 (page 129–131) and 5.3 (page 133–135). I recently developed a set of resources to introduce students to ways of engineering a better understanding of data and statistics through the creation of engaging graph and data visualizations. The projects that are contained in those modules follow the EDP starting with the challenge of creating a story based on data and developing a solution to convey a clear and engaging picture of each story via some sort of graph. The Every Graph Tells a Story project (page 139) is an elementary version of that project. Another mathematics-based activity you can scale down for grades 2–5 is How Does Your Production Line Rate? (page 175). This activity makes the concept of rate and the role of mathematics in planning real to students.

A teacher-favorite project with a wide range of mathematics content is the Building a Better Box project (see figure 5.4, page 136). You can adapt this project to include costs, 2-D layouts to 3-D construction, and area and volume calculations. This project makes students aware of how much packaging materials are used for all sorts of products.

Building a Better Box (Grades 2–5)

This project has two key goals. One goal is to provide an opportunity for students to develop and support spatial reasoning. Strong spatial skills are increasingly tied to student interest and success in STEM fields (Gagnier & Fisher, 2016; Newcombe, 2010; Rule, 2016; Sorby, 2009). The second goal is to bring to light the environmental impacts of packaging, particularly in this age of online shopping and delivery services. This project engages students in the redesign of something all around them. You can structure this project as a short activity, with students spending two or three class periods designing a package for specific contents, or as a longer design project by including minimizing cost and waste, and the creation of 2-D templates that fold into 3-D packages.

Although the project is designed for a mathematics class, you can easily make strong connections to both environmental science and visual arts. In addition, as students test their packages, the concepts related to forces, materials, and structures will be readily evident. Most teachers who have done this project use potato chips as the cargo of choice, but you can substitute any readily available materials that will easily crush or break. You can scale this project for middle school by focusing on creating minimal waste when building the package (doing so requires more planning) and by creating a logo and an eco-friendly marketing campaign.

To maximize this project's value, consider starting by having students look at how much and how ineffectively people use packaging, particularly plastic packaging. According to a report from the World Economic Forum, Ellen MacArthur Foundation, and McKinsey (2016), "95% of plastic packaging material value, or USD 80–120 billion annually, is lost to the economy" (p. 17). A focus on single-use and individually packaged portions has added to the problem. As an article in *Science Advances* (Geyer, Jambeck, & Law, 2017) notes:

> Plastics' largest market is packaging, an application whose growth was accelerated by a global shift from reusable to single-use containers. As a result, the share of plastics in municipal solid waste (by mass) increased from less than 1% in 1960 to more than 10% by 2005 in middle- and high-income countries.

Clearly, this challenge will require some creative engineering, along with consumer awareness. See the project plan in figure 5.4.

Project Title: Building a Better Box	**Topic:** More sustainable packaging; developing spatial-thinking skills
Grade Level: 3–5 Varying levels of visual arts, mathematics content, planning, environmentally friendly design	**Estimated Class Time:** Eight classes

Challenge: Design a package that does the following.
- Holds ten potato chips without crushing or breaking them
- Uses a minimum amount of material
- Uses environmentally friendly materials
- Has visual appeal

Any or all the preceding list can be part of the challenge, depending on time available and desired amount of content.

Curricular Connections	**Skills Focus**
Science—environmental impact of packaging materials; impact-resistant materials; forces **Mathematics**—measurement; 2-D to 3-D planning; economics of efficient planning and material choice (budgeting and costing) **Art**—visual appeal of package design; marketing impact **English Language Arts, English Learners, and World Languages**—any instructions or labeling on the box	**Critical Thinking**—choosing materials; identifying key needs; analyzing effective designs; efficient planning to minimize waste **Creativity**—designing aesthetics; graphic design **Spatial Reasoning**—going from 2-D planning to 3-D models **Collaboration**—working as a team; recognizing common human factors **Communication**—presenting the final package; marketing benefits

Overall Plan	
Hook	**For all grade levels:** Keep a log of the packaging in your home every day. **Grades 4–5:** Watch a video about sustainable packaging design (https://bit.ly/2Jw8c9Q).
Quick Build	Twenty minutes—use limited materials to develop a package that can successfully hold a fully inflated balloon and resist crushing and popping. • Four sheets of printer paper • One 12 × 12-inch piece of corrugated cardboard • Four pipe cleaners • Four popsicle sticks • Twelve inches of duct tape • Scissors Test the box to see how much weight you can place on it before it crushes or the balloon pops! Use textbooks, masses, even students to place a load on the package.
Background Instruction	Forces Materials properties Cost factors Environmental impact Making a template; 2-D design layout
Background Research	Different packaging designs; making observations in the classroom, grocery store, and at home is effective. Packaging materials Different purposes for packaging Understanding recyclable, reusable, and compostable

Engineering Design Process	
Know Your Problem • Know your end user • Identify constraints • Define criteria	• What is the target market for packaged goods? • What cost, size, and safety issues limit the design? • What makes your package design appeal to the consumer and manufacturer?
Know Your Options • Research • Brainstorm	• Further investigate designs if needed. • Use a variety of techniques to get students to consider innovative designs.
Develop a Solution—Part One • Choose a design • Identify needed materials	• Students settle on the best design; students in grades 4 and 5 should discuss how it meets constraints and criteria. • Consider cost or environmental factors when creating a materials list.
Develop a Solution—Part Two • Create a plan (make a sketch) • Build the box	• After some rapid prototyping activities, develop a sketch of the planned package. If required, create a 2-D layout. • Obtain materials to create the package. If students have a budget, stress the need to plan and develop a list. Emphasize planning by not allowing material returns. Reflect the environmental impact with costs (for example, packing peanuts can be very expensive and plain copy paper can be very cheap). Increasing the focus on all these items will increase the challenge.

Figure 5.4: Building a Better Box project plan.

continued ▶

Develop a Solution—Part Three
• Test the box
• Plan some modifications to make the box better
	Depending on grade level, consider the following tests.
	• Consumer feedback: How does it look? Does the package make you want to buy the product?
	• Is it lightweight?
	• Does it resist crushing?
	• Can contents withstand forces due to dropping the package?
	• Can the package hold all of the planned contents?
	• What modifications might you consider?
	   + Does the package need to be more appealing?
	   + Do you need to change materials to protect the contents?
	   + Do you need to change the shape and design of the package to protect the contents?
	   + Is there one thing you can do to make your package better?

Develop a Solution—Part Four • Communicate your results	Create a poster that highlights the following. • The overall package design • The cost of your package • The environmental impact of the package • The results of the testing

Group Size

Grade 3: Introduce groups of three or four. If students have done other projects, consider jobs such as project manager, materials engineer, graphic designer, and marketing manager.

Grades 4–5: Students should begin functioning effectively in teams; jobs are a must.

Assessment

Grade 2:
- Include an assessment component to reflect how well students make and follow a plan.
- The Buck Institue of Education offers a free Creativity and Innovation rubric (https://bit.ly/2xnMNZ8) for this level. Consider using or adapting it.
- Although having individual and group components to grade is unrealistic at this point, emphasize to students that working as a team is important. Formatively assessing and monitoring their teamwork skills is helpful.

Grades 3–5:
- Begin to use a portion of the grade for the project to reflect the need for collaboration and teamwork. This should become a more significant assessment component as students do more projects.
- Identify curricular content understandings to assess along with process components. At this age, students should begin to show understanding of elements of problem definition (end user, constraints, and criteria). There should be evidence of students considering various options. There should also be an increasing focus on ways to test, along with using testing information to make modifications, by grade 5.

Resources

Article titled "What makes a better box?" by Richard Moyer and Susan Everett (2010)

Every Graph Tells a Story (Grades 3–5)

One of the side effects of our technology-rich world is the amount of data generated every day. We will generate an estimated 463 exabytes of data per day (the equivalent of 212,765,957 DVDs daily) by 2025 (Desjardins, 2019). Not all that data gets collated and analyzed, but the amount of information available based on our online habits and communications alone is staggering, and it is often used to predict trends and identify consumer needs and habits. And that is only one major use of all of these data.

But who wants to look through piles of spreadsheets generated by computers? Graphs are much more interesting ways to make data come alive. The world of data visualization is the epitome of a STEAM (science, technology, engineering, arts, and mathematics) project, with a big emphasis on the *M*. The late Hans Rosling was a master at distilling complex ideas down to engaging graphs, both electronically and with simple props. View one of his TED Talks (https://bit.ly/2lpp5nh) for inspiration before trying a project like this.

The Every Graph Tells a Story project focuses on several ideas. See the project plan in figure 5.5.

- It introduces the EDP in a novel way.
- It gets students used to making sense out of a collection of numbers.
- The focus on graphing helps students see how correctly using labels, axes, scales, and formats can tell a story.
- Moving from a 2-D to a 3-D model supports spatial reasoning.
- The strong visual component engages a wide range of learners and the focus on numbers and visuals can foster better inclusion of English learners.

Project Title: Every Graph Tells a Story	**Topic:** Designing a 3-D graph that tells a story about data	
Grade Level: 3–5 Adjusted by the increasing the complexity and level of data as well as the number of required data points		**Estimated Class Time:** Between eight and ten classes
Challenge: Design and build a 3-D graph that can tell a story about lots of numbers. Can your graph tell the story if you are not there to explain it?		
Curricular Connections	**Skills Focus**	
Art—must be visually appealing and moving from two to three dimensions **Mathematics**—graphing skills and relationships between numbers **English Language Arts, Science, Social Studies, World Languages**—can be topic of the graph	**Critical Thinking**—depicting data to convey a message or story; choosing materials **Creativity**—making it stand out; visual appeal to create audience engagement **Spatial Reasoning**—scaling; 3-D thinking **Global View**—using trends and statistics about other regions and countries can be focus **Collaboration**—working as a team; identifying what makes people look at the graph **Communication**—ensuring that people can understand the graph and that it conveys the message	

Figure 5.5: Every Graph Tells a Story project plan.

continued ▶

\multicolumn{2}{c}{**Overall Plan**}	
Hook	Have students explore some innovative graphs online using some from the following resources. • Information Is Beautiful (https://informationisbeautiful.net) • Statista's Chart of the Day (www.statista.com/chartoftheday) • DataVizProject (https://datavizproject.com)
Quick Build	One class period. • Different groups get different subjects to graph based on objects in your classroom or school. • After some initial observations, students list five different features or characteristics that they noticed (five shirt colors, five foods). They make a tally chart to record the incidence of each characteristic. • With little instruction, challenge students to make a graph of their data in about twenty minutes. • Debrief and reflect by having each group show and explain their graph. Other students can ask questions and make comments. • Each group should identify three things that would make their graph easier to understand or more fun to look at. Alternate: Play a graphing game online (Early Childhood Education Zone at https://bit.ly/2XbhpYc).
Background Instruction	• Types of graphs • Labeling • Scale • Legend or key • Any specific related to the graph's subject, if needed
Background Research	Assign or have students choose a general topic for their graph. Topics can be broad such as food, plants, pastimes, and books. Choose anything that relates to your current curricular content if you want to focus on a subject in addition to mathematics. Students find or collect data depending on the topic. If you are introducing large numbers and statistics to students, some good resources may be found on these sites. • Our World in Data (https://ourworldindata.org) • World Resources Institute (www.wri.org/resources) • World Bank data tables (http://wdi.worldbank.org/tables) • NOAA Data in the Classroom (https://dataintheclassroom.noaa.gov)
\multicolumn{2}{c}{**Engineering Design Process**}	
Know Your Problem • Know your end user • Identify constraints • Define criteria	• Students or teachers determine who they show their graphs to: adults, younger or older students, their classmates. • Help students identify constraints; typical constraints include the project's overall size, the time they will have to build the graph, and any materials they may use. • Have groups develop criteria; remind them it must look good and be related to their project. For instance, a project focused on the ocean might have an overall blue color scheme and fish as a theme.

Know Your Options • Research • Brainstorm	• Help students identify a good data set related to their topic. • Brainstorming should focus on ideas like the type of graph, colors, and objects that would make it 3-D. Lead brainstorming by give periodic prompts related to these ideas.
Develop a Solution—Part One • Choose a design • Identify needed materials	• Just as they would in any EDP projects, have students decide or vote on key features of their graph, including type, color, theme, objects, and materials. • Students create an initial list of materials.
Develop a Solution—Part Two • Create a plan; make a sketch • Build the graph	• Visit **go.SolutionTree.com/21stcenturyskills** for an abbreviated "Initial Design Plan Graph Project." It is important that students complete this before starting to build. Jobs are listed but they are optional for this project. • Allow three or four classes for building; remind students that graphs must not be flat (2-D), must have the key features you taught them about (labels, units, scale, and so on), and must meet the constraints and criteria.
Develop a Solution—Part Three • Present the graph • Get feedback • Plan modifications to make it better	• Groups present their graphs in a gallery-like setting where you invite others to come and view them in your classroom. An extension or added challenge has students be absent during this visit, since the graph should be able to tell the story. • Have groups create a three- or four-question Likert scale survey to give to viewers. Questions should include whether the story is clear and about color and overall theme and presentation. Students must understand that they are looking for information about how clear and engaging their graph is. • Once the galley tour is over, guide groups in analyzing their surveys. Ask each group to list three ways they could improve their graphs.
Group Size	
Groups of three work well; if necessary, groups of four will work.	
Assessment	
Follow the general assessment guidelines in chapter 3 (page 80–91).	

*Visit **go.SolutionTree.com/21stcenturyskills** for a free reproducible version of this figure.*

STEAM-Based Projects

In grades K–5, looking to both nature and the visual arts can help you develop highly engaging projects that connect well to what you already teach. At its heart, engineering is all about problem solving and finding innovative solutions to meet human needs and improve lives. Good engineering demands an understanding of systems impacts, trade-offs, and unintended consequences. Nature inherently operates on those principles, taking a long-term, sustainable approach to optimization—and nature has been successfully engineering for about 4.5 billion years. As Janine M. Benyus (1997), a champion of biomimicry, says, "The more our world functions like the natural world, the more likely we are to endure on this home that is ours, but not ours alone" (p. 3).

Biomimicry is the practice of observing and emulating nature when designing and engineering solutions. It provides a terrific platform for incorporating engineering design projects into life science, environmental science, and biology curricula. Looking to nature to model sustainable engineering design practices engages students. In addition, using biomimicry or bio-inspired engineering addresses all sorts of real-world problems—from energy production to clothing design.

Before beginning a biomimicry-based project, you and your students may want to explore how nature deals with the constraints of localized resources and simple criteria related to sustaining life. Contrast that with our human ability to secure resources from all distances, and our drive to make life better with often-changing criteria. Nature's approach to modifying and optimizing is evident in the adaptation and survival of the fittest, a slow process based on optimization of resources and minimization of negative impacts. Human design often focuses on maximizing desired qualities or features. Our rapid iteration process rarely allows enough time to assess the full impact of our solutions. Table 5.1 compares and summarizes both approaches.

Table 5.1: Nature Versus Human Design

Engineering Design Process	Nature's Approach	Human Approach
Define the Problem	Sustain life	Make life "better"
Identify Criteria	Nontoxic, low temperature, recyclable, renewable	Bigger, stronger, cheaper, safer, appeals to target audience
Determine Constraints	Only locally available resources	Least-expensive resources, limited time
Testing and Modification	Slowly over time, prolonged use, extensive population	Rapid, often limited scenarios, small pilot samples
Optimization	Sustain life with least negative impacts; positive impacts outweigh negative consequences	Bigger, better, faster, more profitable; often maximize not optimize

The following projects work well with students at all grade levels (K–12). In addition, some modifications and extensions can help you create multiple projects or projects that allow for differentiation in your classroom.

Hidden in Plain Sight Project—Biomimicry (Grades K–2 and 3–5)

This challenge asks students to take inspiration from the colors and patterns in nature to design a better butterfly (grades K–2) or camouflage for a nature photographer (grades 3–5). Students explore patterns and designs in nature while learning about adaptation and natural selection. The final prototype is either a butterfly designed to hide in the classroom "garden" or a T-shirt designed to hide the photographer in a specific environment. Along the way, students learn about different types of camouflage and patterns in nature. Their final challenge is to put some of this knowledge together to create their camouflage artifact.

Both projects can start with a simple hunting activity that involves students finding various colors and shapes of creatures on a wrapping or scrapbook paper background. Visit **go.SolutionTree.com/21stcenturyskills** to access the instructions for these two versions.

Since there is a wide range of possibilities in this project, table 5.2 summarizes how it might flow and fit the EDP.

Table 5.2: Hidden in Plain Sight Project Summary

	Better Butterflies	**Photographer Camouflage**
Initial Hook	Reading *The Very Hungry Caterpillar* (Carle, 1994) or *A Butterfly Is Patient* (Aston & Long, 2011)	Pictures or videos about Liu Bolin, a Chinese artist who is known as the Invisible Man (https://bit.ly/2J3dHfx). He camouflages himself in an amazing range of scenes and places.
Initial Engagement Activity	Butterfly-hunting activity	How does nature design camouflage? The Create-a-Critter activity challenges students to make small "critters" (maximum size is 2 x 3 inches) to hide in various parts of the school library or cafeteria. Encourage students to use different patterns and types of camouflage (concealing coloration, disruptive coloration, disguise based on shape or texture, mimicry, and so on). Project Learning Tree (2017) has some great examples, as do many other websites.
EDP Phase 1: Know Your Problem Define problem; identify constraints and criteria	Define the problem as "survival"—what colors and patterns hide the butterfly so it can survive? Constraints—materials on hand (same as nature) Criteria—matches with the environment	Problem is the need to conceal the photographer, often while he or she moves in the environment. Constraints—materials on hand Criteria—some sort of match with the environment, plus comfort and climate suitability

continued →

	Better Butterflies	**Photographer Camouflage**
EDP Phase 2: Know Your Options Research what works; examples of different camouflage in nature and man-made Brainstorm possible solutions; focus on color and patterns	Students may also do some exploring outside in this stage.	It may be helpful for individual students to create small sample patterns as part of or subsequent to brainstorming. They should focus on the idea of quantity over quality before settling on a solution.
EDP Phase 3: Develop a Solution Make a prototype; test; modify; communicate results	Have students test their butterfly by letting a few different students hunt for it while the designers record how long it takes for someone to find it. They can also obtain feedback on what gave their butterfly away. Either allow time for modifications or ask students what modifications (same as adaptations) would help.	Test T-shirt designs against the environment picture or on a green screen with the environment picture projected. You may even want to involve students in a do-it-yourself green-screen project or in the use of Photoshop to achieve the effect. The most effective testing is based on group feedback on what is easily noticed, for example. Allow modifications if there is time; if not, ask for a list of two or three possible modifications. Students present their findings as marketing pitches or posters for *National Geographic*. Have students highlight the best features of their design and include some accessories to complete the look. Students should also identify specific regions of the world that would suit their camouflage design.

Visit **go.SolutionTree.com/21stcenturyskills** *for a free reproducible version of this table.*

Butterfly

For the butterfly version (grades K–2), the teacher uses a traditional hole punch or, for a more engaging approach, a butterfly-shaped hole punch (available at craft stores) to create various color paper butterflies. Using scrapbook or wrapping paper with different floral or other natural designs, have students predict what colors will "survive" in different environments. Sprinkle some creature shapes in different environments and give students a few minutes to hunt in low light, which simulates the dawn or dusk lighting most predators prefer.

Students can count the number of each color that they found and, if you would like to include a mathematical component to your project, students can show the results in bar graphs. They can create a colored bar graph. For instance, if they found five yellow butterflies, they can color in a vertical bar five units high on a graph, and then do the same for all the other colors. You can discuss, as a class, why certain colors help butterflies hide, and why others may not work as well. Nature *modifies* the butterflies through a long-term process of adaptation that favors the ones that are better hidden. Different colors and patterns work better in different environments.

Depending on your students' grade level and abilities, you can talk about symmetry and Fibonacci spirals in natural design. You can search online for *Fibonacci elementary* and find ideas (such as that at www.mensaforkids.org/teach/lesson-plans/fabulous-fibonacci). A *Fibonacci* sequence starts with 0 and 1, and then each number after is the two previous numbers added together. It is depicted in a spiral like you see in a nautilus shell and the seeds in the center of a sunflower.

T-Shirt

In the T-shirt version, give different groups pictures of different environments. As students begin to design a T-shirt for their environment, provide helpful resources and pictures. The Biomimicry Institute website (https://biomimicry.org) and its companion website Ask Nature (https://asknature.org) have a wide range of bio-inspired design resources and photographs.

Abby Sutton, an eighth-grade STEAM teacher at Alan Shawn Feinstein Middle School in Rhode Island, says:

> After teaching a traditional science class for many years, I was given the opportunity to teach an eighth-grade STEAM class that focuses on life science concepts. This has been an incredible learning experience for me as an educator. I have observed my students embrace failure and work collaboratively to tackle each design challenge. I have been amazed at their level of engagement, creativity, and ownership over their design solutions, and I have realized that they have not only learned about the engineering design process and biomimicry throughout this course they have also learned to take risks, ask questions, work creatively together, investigate ideas, design innovative solutions to real-world problems, and communicate results with one another. In addition to learning science content, my students have developed lifelong skills that will be critical to their success in college and the workplace. (personal communication, June 30, 2018)

No Words Activity: Pictorial Instructions (Grades K–5)

This is a short design activity I often use as background for more involved projects that require the designers to create instructions. However, I find many teachers like to use it as a stand-alone activity. This activity will help you jump into using engineering design practices. It is also a wonderfully inclusive activity for the ELs in your classroom. I have seen it used at all levels (preK–12), and it can be a great team-building exercise early in the year.

Pictorial instructions appear in many places, from Ikea furniture to computers. In general, you can convey a lot of information with well-designed picture sequences. From an engineering standpoint, pictorial instructions provide a way to convey ideas to users who have limited literacy, as well as eliminate the need and expense for countless translations. Developing pictorial instructions for assembly or a process requires sequential planning, clarity, awareness of the target audience, and, often, a degree of spatial reasoning. These skills have value in both STEM and non-STEM fields.

Any of the No Words challenges I use start with a simple pictorial instructions example. Find one that works for your class by searching online for *pictorial instructions* and clicking on Images. I challenge pairs of students to create some instructions for tasks such as brushing your teeth or making a sandwich. Sharing the instructions with each other illustrates the value of feedback and working with an end user.

The actual design challenge involves LEGOs (or DUPLOs or larger bricks). In general, this activity takes about one hour. These are the directions.

1. Give groups of three to four students two identical sets of bricks, each in a plastic bag. The students create a sculpture with one set and give it a name.

2. Groups develop assembly instructions that only contain *colors, shapes, arrows,* and *numbers*—no words allowed!

3. Groups hide their sculpture and put their instructions in the remaining bag.

4. Groups switch instructions and building sets with other groups and then attempt to recreate the sculptures. Allow time for students to share their experiences and instruct them to think about what was helpful and what was confusing.

You can adjust the challenge level by increasing the number of pieces. For students in preK–2, between eight and twelve DUPLOs or other large interlocking building blocks are generally a good number. For grades 3 and above, I generally use between twelve and fifteen LEGOs or similar interlocking building blocks. For students in grades 6 and above, you can add to the challenge by instructing groups *not* to use one or two pieces in their original sculpture but avoid indicating which pieces were left out. The group attempting to build their sculpture will then have the added task of identifying the outliers.

Most teachers who try this activity remark on its team-building aspects. In addition, it highlights engineering skills such as planning, communication, and the value of feedback for improving a solution. You can add a mathematical component to it by requiring scale and graph paper. Middle schoolers can begin to understand the idea of subassemblies in manufacturing; you can require them to make their sculpture with three subassemblies, each with its own instructions, along with a final assembly procedure.

Going Forward

The key thing to remember for any project is to frame it with the EDP. This gives you a way to easily incorporate 21st century skills and real-world connections. In addition, increased exposure will make the EDP part of your classroom routine. It will eliminate the need to instruct students about how to follow a procedure to develop solutions to somewhat messy problems, allowing students time to explore curricular connections and apply them to meet the design challenge. Students can focus on the *what* and *why* more if the *how* becomes second nature. The opportunity to think, apply, notice, and care are those ounces of action that will help students transfer their understanding to the future beyond your classroom. And, as Friedrich Engels notes (as cited in Andrews, 1990, p. 2), those ounces of action may have more value than the tons of theory that have been the focal point of education.

CHAPTER 6
Introducing Projects for Middle and High School

What you know is far less important than what you can do with what you know.

—Tony Wagner

If you are a middle or high school teacher, chances are you have complained about how much content you are required to teach. Science, in particular, seems to be the focus of an ongoing debate concerning what topics teachers should teach and in how much detail. History curricula often become the subject of similar debates. The debate over breadth and depth is not new. The idea of exposing students to new ideas and preparing them for the future has always been at the heart of education. But the explosion in knowledge and information over the past one hundred to one hundred fifty years makes the task of imparting all known facts impossible. Newer initiatives and standards, such as the NGSS, stress the importance of core ideas or key understandings in a wide range of disciplines (NGSS Lead States, 2013). Yet, in many cases, textbooks and large-scale testing still embrace a broad survey approach. It is enough to make any teacher's head spin!

Most of the projects in this chapter work for grades 6–12. You can customize the depth of content knowledge or complexity for many of these projects. For example, the prosthetic hand challenge can be less complex if students focus on how their own hands work and take a more superficial view than they would in a high school anatomy class, where you might connect some of the underlying anatomical structures to the project. I have even seen that project add more layers through the use of 3-D printers to make various components and servomotors to make the hand move. It all depends on your expertise, your learning goals, your students' ages, and your resources. I am a huge fan of water footprint projects and have seen a wide range of understanding and final project deliverables. You can give students the simpler task of creating awareness by developing simple poster presentations of a product's overall water footprint, or they can create videos showing the water use component of various stages in a product's life cycle.

In my experience, adjusting those two areas (depth of content knowledge or complexity) will allow you to scale most middle and high school projects appropriately. In many cases, adjusting the complexity of the background content or the expected prototype can allow teachers in grades 3–5 to use modified versions of these projects or to plan similar ones. The board game project is often used in these grades with adjustments for designing packaging and a bit more assistance as students try to write instructions.

I chose these projects intentionally for their adaptability at different grade levels and across a range of disciplines. Most have added dimensions based on strong visual arts, social studies, or mathematics components. This helps you engage student interests and talents, taking the focus away from highly technical skills. It can also help you involve other faculty members in your project. There are some parallels to chapter 5 (page 125) projects, so borrow and adapt activities and planning ideas from some of those plans and discussions. Remember that, whether we are eight years old or eighty, we all engineer!

Best Practices for Middle and High School Students

Starting small at grades 6–12 may be more critical than it is with students in grades 5 and lower. Curricula jam-packed with concepts can make it challenging for teachers to include active and student-led learning in lesson plans. Furthermore, grades 6–12 students have mostly mastered the game of school and may be less willing to fail or consider there may not be one right answer. Use some shorter activities and projects to cultivate and enlist students' enthusiasm and engagement before jumping into larger-scale projects. Make failure an acceptable option and the possibility of multiple solutions part of your classroom culture before embedding these ideas into a curricular project. Also, remember these are general project overviews; they are not complete project or lesson plans, so modify them to fit your classroom and curriculum, but keep the process in the forefront.

Based on my work with teachers at these grade levels, I suggest a couple things. Don't assume that students know how to work collaboratively in groups. They may have had experience in group work by this age, but it is often a one-class phenomenon and may not have involved managing different roles and steps. Take some time to review the Increasing Synergy with Teamwork section (page 70) in chapter 3 before launching your first project. Doing some shorter activities from chapter 4 (page 95) and watching various group dynamics is helpful. You may also want to consider a holding a Teamwork 101 session beforehand to have an open discussion of roles, practices, and goals for working this way. Remember that if we expect students to learn how to collaborate, we should expect to teach them how to do it.

Also, address student concerns about how working in groups will impact their grades. They are justified in this and you need to address it. How assessments translate to grades should not be a secret. Their world is often governed by the letters on their report cards. And, for the most part, they are masters of their own ships when that grade is developed. Clarify that you have safeguards in place to minimize the impact of a disengaged or nonproductive group member. Peer review, your ongoing classroom observations, and the responsibilities of various roles help you here. Consider having procedures in place to manage any issues where one student is not contributing or is negatively impacting the group. Group conferences work well as a first step, followed by conferences with the individual if needed. Consensus on reassigning responsibilities or some milestones for specific contributions may be all you need to change the dynamic. But it is important that your students know that as they are each learning how to collaborate, their grades will not be impacted by the one student who has opted not to be a productive member.

Overall Approach

Many of the projects in this chapter support concepts from more than one discipline. How well that works depends very much on your school and your colleagues. It is difficult to model real-world problem solving and design in isolated subject silos. Any effort beyond your classroom walls will send a strong message to your students, so please try to plan for some support. Enlist the support of the administrators in your building by inviting them to visit your classroom during the building stage. That is the point when most students are eager to share what they are doing and why they are doing it. Ask to speak at department and faculty meetings about what you are trying to do; you might be able to recruit some of the expertise on your own faculty. Extend an open invitation to other teachers to stop by and see things in action.

Let your students know if a colleague who has knowledge of or firsthand experience with something related to your project is willing to be interviewed. Take your projects and your students on "field trips" to other classrooms to share ideas or to make some observations about potential end users. Intentionally make the message very clear to your students that schools are amazing places full of all sorts of talents and expertise, and take action by including those resources in your projects. You will have to be strategic in the face of varying student schedules, different teaching practices and beliefs, and the ever-present concern with extensive coverage that frames many curricula. It is often easier to manage multidisciplinary projects in middle schools with some team teaching.

If your school is strongly divided by departments (as many high schools are), consider tapping colleagues as experts in subjects beyond your discipline. Colleagues can help you with planning or serve as guest lecturers or audience members for students sharing their solutions in your classroom. The more you involve your fellow faculty members, the more likely they are to want to learn more. In addition, you are modeling the collaboration you expect from your students. Schools that function as professional learning communities are at an advantage here because they already work collaboratively.

> Colleagues can help you with planning or serve as guest lecturers or audience members for students sharing their solutions in your classroom.

STEAM Up Front

In a reversal of the chapter 5 layout and in the spirit of crossing some of those disciplinary boundaries, I'll lead with STEAM-related projects in this chapter. These projects are great ways to engage students who have opted out of traditional STEM classes and identify as artists. The world needs them! Their talents and perspectives add to any design group and create a richer tapestry of possible solutions to the challenges ahead. Much of my initial work was aimed at re-engaging these students. They generally bring strong spatial thinking and the studio-thinking habits of envisioning, observing, and reflecting to the challenge (Project Zero, 2003). They are also more open to unique approaches and feedback or critique.

Unfortunately, many students who identify as being creative lose interest in STEM fields during high school because of a tendency to view them as uncreative endeavors. Furthermore, this belief that science and mathematics are not creative subjects is more prevalent in girls (Valenti, Masnick, Cox, & Osman, 2016). There is no need for everyone to become scientists or engineers, but we all benefit by including as many perspectives and talents as possible in the development of the designed world. To do that, we must encourage interest at an early age.

You can easily adapt both of the following projects to any middle or high school grade level. Teachers use both in a wide range of classrooms and they generally bring back these projects for return engagements in subsequent years. If you are looking for an engineering design project to use in *any* subject, consider the Engineering a Board Game project; if you are looking for projects that combine visual impact and systems thinking, the Water Footprint projects work well in mathematics and life science classes. The implications of water movement across country boundaries can even be the basis for an interesting strategy game in social studies classes.

Engineering a Board Game—Visual Arts and Communication (Grades 6–12)

University of Wisconsin-Madison professor Constance Steinkuehler, an expert in using games in education, points out, "Games are architectures for engagement" (as cited in MacKay, 2013). Our classrooms should always be places where students are engaged in learning. Why not combine the fun of games with an engineering design challenge? Think of using a board game design project as the way to introduce the EDP into *any* classroom or subject. You'll be engaging students in a creative and collaborative project that involves planning, spatial reasoning, and technical writing, and providing an opportunity to apply and work with concepts from *any* topic.

Board game design is the ultimate STEAM project, requiring content knowledge, spatial and mathematical reasoning, visual design components, and the ability to communicate clear instructions. This project meets NGSS, Common Core, and a range of discipline-specific standards, and you can scale the project to meet different grade-level and class time requirements. Game and toy design are also valuable and highly creative engineering design experiences. There are even college-level engineering courses based on toy design projects (Greim, 2010).

Board game design is the ultimate STEAM project.

A great deal of engineering goes into developing a board game. Engineers must design packaging, create themes and layouts, develop and describe the playing process, source needed materials, test and modify the game by obtaining user feedback, and present the game in an engaging manner. Add the ability to focus the theme on any curricular topic or unit and you have an amazing platform to engage the future toy engineers or game designers in your classroom in a unique learning experience rich in both skills and content. You can easily modify this project to support differentiated learning and, due to the strong visual component and the ability to shift the project requirements to different perspectives, you can adapt it for use in English as a new language and other EL classes. I suggest this project for world language classes since it is a great way to explore vocabulary, culture, customs, and geography.

The entire project highlights and centers on the EDP. Different topic areas and complexity levels allow you to customize the general project for a wide range of ages and grade levels. Some teachers have added requirements for sound effects or a light triggered by a certain play or move. These have included simple switch-activated circuits and Arduino-controlled effects. In schools with the technology, students print 3-D game pieces. A handful of teachers have used this project as an introductory activity to develop an initial prototype prior to electronic game design and coding projects. Resources supporting learning from making and playing games are listed in the project plan.

Since this project is easily and often modified and customized, there are some things you should consider in terms of planning your project. The first involves content or topic choice.

- Choose your topic and decide how many concepts or ideas you want the game to feature. For example, if your middle school students are learning about cells, you may want to provide a list of fifteen to twenty key facts or terms, and require at least ten in the game. Some topic ideas follow.
  - Any core science topic or unit
  - Sets of mathematics facts
  - Shapes and geometry concepts
  - Global issues, such as types of renewable energy, plastics in the ocean, various products' water footprints, the United Nations' (n.d.) sustainable development goals
  - World language games can focus on key vocabulary, phrases, and cultural issues
  - Games in English language arts classes can revolve around characters and plots in stories and novels
  - History games about important dates and events; cultural and economic aspects of development
- Start with a clear description of the content you want to feature in the game. I highly recommend providing students with a list of the specific topics or concepts.
- All groups can focus on the same topic since they will all develop different games based on their own criteria and research. Or, you can maximize the exposure to different aspects of a topic by assigning different subtopics to different groups.

The second thing you should think carefully about relates to developing constraints and required features.

- Provide clear constraints for your students. You will probably want to give them an overall footprint (length times width) for their board and some guidelines for the size of their box or package. (Some of the middle school teachers I work with provide a box groups can customize for their game; this does save some time in the project.)

- Make packaging requirements smaller than the board requires so some engineering about how to fold the board is necessary.
- Make sure you are clear about any materials requirements. Remember, at a minimum, games will involve a board, playing pieces, some sort of method for determining where or how far pieces move, and a box. Many games will also include cards, fake money, and prizes.
- Be very clear about the instructions format. If you want a particular layout, clearly specify it, and consider providing a template or developing one with your class.
- Consider requiring a name for the game and some sort of branding or logo. Remind students that the name generally reflects the theme.
- Specify an age range of players. Ideally, you should bring in a group of testers who fit the demographic for feedback. Most of the successful designs I have seen aim for a slightly younger age group than the designers. But I have also seen advanced placement (AP) students challenged by a middle school–designed game!

This project has a number of components and requires students do some planning. It is critical for students to have jobs so they can divide and conquer—or work on different tasks concurrently. Table 3.2 (page 74) shows the guidelines most teachers follow to set up jobs. Part of your planning requires having materials on hand.

The project plan in figure 6.1 is more detailed than most in this book because teachers from several different disciplines often use this project, and some may be less familiar with the EDP. This is a popular project because it relates to something students know about and have used, and because it involves a lot of visual arts and planning skills. Simplify it as needed, but try to keep the design process as intact as possible. Figure 6.2 (page 158) shows the components and materials plan checklist, and figure 6.3 (page 159) shows the packaging design plan.

Project Title: Engineering a Board Game	**Topic:** Designing an educational board game that engages players as they learn about a topic
Grade Level: 6–12 Varying levels of game difficulty, prototyping requirements, and topic or subject content including requiring more content inclusion, adding electronic effects, and more complex rules and including instructions will scale this project from the middle to high school level.	**Estimated Class Time:** Ten to fifteen classes, depending on the grade level
Challenge: Create a board game to help students and others learn about a subject or topic. Note: Teachers should specify age level of players and required content. Project content includes visual layout and appeal, technical writing for instructions, and the value of consumer feedback. You can also use this project to focus on planning and collaboration skills due to its multifaceted nature.	

Curricular Connections	Skills Focus
Mathematics—measurement; 2-D to 3-D planning; analysis of Likert scale feedback data **Art**—aesthetics of packaging and board design **ELA**—technical writing of instructions; any trivia-type questions part of the actual game **All others**—Topic or subject focus of the game connects to any chosen discipline	**Critical Thinking**—choosing materials; identifying key needs; analyzing effective designs **Creativity**—deciding design aesthetics **Spatial Reasoning**—going from 2-D planning to 3-D models; human interaction with designed environments and objects **Collaboration**—working as a team; recognizing common human factors; managing a project with many aspects (board design, packaging design, strategy, and marketing) **Communication**—listening to and acting on consumer feedback; presenting the final game; creating clear instructions and questions
Overall Plan	
Hook	There are a number of short videos and articles about the game Monopoly's educational nature. Search *story behind Monopoly* to find one at an appropriate level for your class. • Watch "The Surprising History Behind the Board Game 'Monopoly'" (www.youtube.com/watch?v=mz5H0cg2uXs). • Watch "This Woman Invented Monopoly to Combat Greed" video about the history of Monopoly (https://bit.ly/2XBQQzL).
Engagement Activity or Quick Build	• Each student interviews five people about their favorite board game. • Develop questions (similar to pain-point interviewing) with the entire class; have students focus on what they think they need to know to make the game fun and engaging. Create a graphic organizer to record interviewee responses. • Allow class time for group members to compare and analyze responses.
Background Instruction	• Use any content-related (theme) information students need to understand and include in the game. • Consider including an activity where students write simple step-by-step instructions for an everyday activity (brushing your teeth or making a sandwich, for example) as preparation for writing the game instructions.
Background Research	• Use additional content background, particularly connections and trivia-type information. • Research different game designs and strategies. • Investigate best ways to build a box. Note: Some of the ideas and resources from the Building a Better Box project (page 135) might be helpful.

Figure 6.1: Engineering a Board Game project plan.

continued ▶

\	**Engineering Design Process**
Know Your Problem • Know your end user • Identify constraints • Define criteria	• Think about the given age group; survey the types of games they play; how well they can read; and explain concepts clearly for a particular age group. • Teacher-supplied constraints include board size and number of players. • Use research about the end user to develop a list of features players like; discuss color and packaging specifics as possible criteria.
Know Your Options • Research • Brainstorm	• Remind students they need to know concepts well to explain them to younger students via the game; encourage students to investigate strategies in some classic games. • Use a variety of techniques to get students to consider innovative designs; remind them to consider layout, color themes, design and packaging, name, player pieces, and so on.

Develop a Solution—Part One
• Choose a design
• Identify needed materials
	• Students settle on the best design and discuss how it meets constraints and criteria.
	• Encourage students to keep it simple; see the suggested materials. Reusing and repurposing materials such as old games and toys from home is fine with teacher approval. Students need to create and build their own board and packaging.
	• The following common materials work for this project.
	    + Chart or craft paper (for planning)
	    + Foam core (for game pieces)
	    + Cardboard (recycled, for boxes and boards)
	    + Card stock (various colors, for any cards students may wish to create)
	    + Buttons, pom-poms, pipe cleaners, and so on (for making and decorating game pieces)
	    + Scissors
	    + Transparent adhesive tape
	    + Duct tape (various colors)
	    + White or craft glue
	    + Hot glue guns and glue (optional)
	    + Student-supplied materials (recycled, not new; with teacher approval)

Develop a Solution—Part Two
• Create a plan; make a sketch
	• Students need to plan the following before building anything for this project.
	    + Game components (see figure 6.2, page 158)
	    + Game concepts based on your requirements. It is critical for students to list these; some teachers also request information about where each appears in the game.
	    + Game packaging (see figure 6.3, page 159)
	    + Instructions (can be done concurrently with building but a general plan should be in place beforehand). Encourage students to look at actual game instructions. They need to think about how pieces move, penalties, rewards, initial set-up, what determines winning, and so on.
	• Obtain materials and build the game, pieces, and box; keep material choice simple.
	• Work on detailing instructions.

Develop a Solution—Part Three
• Get feedback
• Plan modifications to make it better
	• Test using consumer feedback.
	    + Find a good audience.
	    + Consider having test groups rotate through two or three games to get enough feedback (you may need to limit play time to ten to twenty minutes).
	    + Work with students before testing to develop Likert scale (1–5) surveys for testers. Five questions and a comment space work well. Encourage students to consider their criteria when they develop questions.
	    + Have testers set up and play the game without input from the student engineers. If necessary, students can provide a prompt and should note what was unclear.
	    + Encourage your students to take notes as they observe how testers play the game.
	    + Allow one or two minutes at the end for testers to complete the survey.
	• Try modifications.
	    + Students must connect their modifications to their testing information (observations or survey data). Create a very simple modification form for this task. The form needs to ask what the modification is and why it is being done.
	    + Limit modifications or any need to plan and follow a design process will be lost. Two or three modifications are generally enough.

Develop a Solution—Part Four • Communicate your results	• Since this is a consumer product, a marketing pitch or ad campaign works well and eliminates the opportunity for a rote and boring presentation. • Limit time and give students a list of a few things that should be in their pitch. • Have the students in the audience rate or rank the pitches, just for fun.

continued ▶

Group Size	
Four or five students should begin to function effectively in teams; jobs are a must because of the number of planning and building steps. Indicate some suggested jobs in the project planning discussion. Alter the jobs as needed (see table 3.2, page 74), but look for a diverse and equitable distribution of responsibilities.	
Assessment	
Follow the general assessment guidelines in chapter 3 (page 80–91). Key considerations specific to this project follow. • Although it is a group project, many tasks are done individually and concurrently. The individual component of the grade can be a bit higher (30 to 40 percent) than I normally suggest, and it can focus on jobs. • Other useful tools include a simple teacher checklist indicating who is on task every day with an opportunity for peer assessment. • Some teachers opt to include a short quiz about the concepts and focus of the game, which becomes part of the individual assessment.	
Resources	
• "Toying With Education" (https://bit.ly/2JarulD) • "Board Game Helps Mexican Coffee Farmers Grasp Complex Ecological Interactions" (https://bit.ly/2WyC1tW) • "Playing to Learn: Panelists at Stanford Discussion Say Using Games as an Educational Tool Provides Opportunities for Deeper Learning" (https://stanford.io/2VU9Vw4)	

Visit **go.SolutionTree.com/21stcenturyskills** *for a free reproducible version of this figure.*

Group members:

Directions: Please complete the following checklist to help you plan your game and gather the materials. Attach sketches of your board, cards, and pieces. Add any parts you will use to the list.

Part	Number	Description	Materials
Board			
Players' pieces			
Fake money			
Cards			
Spinner or dice			

Figure 6.2: Engineering a Board Game—components and materials plan checklist.

Visit **go.SolutionTree.com/21stcenturyskills** *for a free reproducible version of this figure.*

Group members:
Possible game names: 1. 2. 3. **Game type (circle one):**   Card   Board   Other **Overall shape of box or outer packaging:** **Additional storage components (to hold small pieces or cards):** **Materials:** **Please attach a sketch of your planned game box or package.**

Figure 6.3: Engineering a Board Game—packaging design plan.

Visit *go.SolutionTree.com/21stcenturyskills* for a free reproducible version of this figure.

Where Does All the Water Go? Water Footprint Awareness (Grades 8–12)

Students and teachers alike find the concept of the water footprint (WFP) of goods and services fascinating. The WFP creates a terrific engineering design project foundation that centers around the idea of making people aware of an issue; it stresses that creating awareness is often the first step in solving a problem.

Statistics concerning actual daily personal household water use (taking showers, flushing the toilet, washing laundry, and so on) range from sixty to ninety gallons per person in the United States, with most people estimating they use much less (Attari, 2014). Our understanding of how much water we use is further complicated because most of it is hidden in the production, transportation, and disposal of consumer products and services. This is known as your *water footprint*. Estimates of the average daily WFP in the United States are generally around 2,100 gallons per day (Water Footprint Network, n.d.a). Reducing what you use directly to shower, wash clothes, brush your teeth, and so on is helpful, but clearly not the biggest factor in conserving water. What you eat, the energy and the transportation you use, and the products you purchase account for a far bigger percentage of your impact, so changes in your choices and lifestyle can go a long way in furthering conservation.

The idea of the WFP engages students (and teachers) because of the sheer size and impact of the numbers. For instance, one can of soda has a WFP of 124 liters (33 gallons); an egg has 200 liters (53 gallons), and a cup of coffee has 140 liters (37 gallons; Hoekstra & van Heek, 2017). Levi Strauss (2015) conducted a life-cycle analysis of a pair of 501 jeans and calculated a WFP of 3,781 liters

(1,000 gallons)! This project also provides the perfect opportunity for students to use conversions or to develop a better sense of metric units. Graphic comparisons to bottles of water and full bath tubs or swimming pools help to bring these numbers home.

There is a lot of content behind the WFP. Since 2002, researchers have based actual measurements of the WFP on a well-developed, complex, refined, and verified mathematical model (Hoekstra, Chapagain, Aldaya, & Mekonnen, 2011). Complete WFP analyses involve blue, green, and gray water consumption. *Blue water* is surface water, *green water* reflects precipitation stored in soil and plants, and *gray water* measures how much clean water must be added to polluted water to make it safe. All this supports rich content connections in mathematics, earth science, and environmental science classes. In addition, the movement of water across country borders due to its inclusion in goods and services means countries are either net importers or exporters of water. Industrialized countries have significantly higher WFPs per capita than most of the world's poorest countries; this lays the foundation for connections in social studies classes (Hoekstra et al., 2011).

This project is essentially about engineering public awareness. The final product can be a video, website, board or electronic game, or a gallery or guide highlighting WFP impact and management. Figure 6.4 shows a general project plan checklist, but there can be a lot of variety in terms of project focus and structure.

Project Title: Where Does All the Water Go? **Topic:** Investigating both (1) how much water we really use and (2) ways to design platforms to increase awareness; project content includes use of graphics and visuals for impact, understanding of data and statistics, environmental and economic impact of water use; strong critical-thinking and communication components; high degree of systems thinking.	

Grade Level: 8–12
 Modify based on:
 • Understanding and use of water footprint data and statistics
 • Water use, water cycle, and environmental content impact levels
 • Product deliverable
    + Video   + Game
    + Website   + Infographic
    + Gallery
	Estimated class time: Five to fifteen classes, depending on grade level and content

Challenge: Engineer a way to increase awareness of the water footprint. The teacher should specify acceptable type of product (platform) and required content.	

Curricular Connections	Skills Focus
Mathematics—data, statistics, modeling, and graphing; analysis of Likert scale feedback data **Art**—aesthetics and planning of display, graph, website, or video **Science**—water use; water cycle; environmental impacts; production of goods, energy, and transportation	**Critical Thinking**—choosing materials; identifying what makes an impact; analyzing effective designs **Creativity**—aesthetics of design; visual impact **Collaboration**—working as a team; recognizing common human factors; interpretation of data and visuals **Communication**—observing what generates engagement and understanding; listening to and acting on consumer feedback; presentation of final project **Systems Thinking**—understanding inputs, outputs, impacts, and unintended consequences

	Overall Plan	
Hook	World Wildlife Fund-Canada (www.wwf.ca) has a short animated video that provides a good overview (www.youtube.com/watch?v=0_bUzH6T6zU). Find others on the Water Footprint Network site under Resources (https://waterfootprint.org/en).	
Engagement Activity or Quick Build	Have students develop an estimated personal water footprint for just the foods they eat. (You can also do this for one meal.) The Water Footprint Network and the Water Footprint Calculator product galleries are helpful. (See Resources.) Allow class time for group members to compare and analyze responses.	
Background Instruction	Provide a brief overview of the water footprint; blue, green, and gray water; specific content you want to reflect in the project; any content-related (theme) information students will need to understand to include in the game. The Water Footprint Network (n.d.b) has a well-designed presentation (https://bit.ly/2PSB2lR).	
Background Research	Research water footprints of different products and processes. Categorize them (such as agriculture, clothing, energy, or transportation). Identify the most water-intensive parts of their production and use. This gives students an idea of water use in a wide range of products and possibilities for decreasing consumption. It is helpful for getting students to think about and create awareness of possible solutions.	
	Engineering Design Process	
Know Your Problem • Know your end user • Identify constraints • Define criteria	• Think about the target audience. What types of statistics and images will really grab them? • Constraints are mostly teacher-supplied, depending on desired end product. • Stress the importance of a central message or vision to help students determine criteria. • Use end user research to develop a list of features important to convey information. Discuss color as a possible criterion, and the overall project themes and layout.	
Know Your Options • Research • Brainstorm	• Remind students to keep their message in mind when researching; encourage students to investigate how different professionals design for visual engagement and impact. • Use a variety of techniques to get students to consider innovative designs; remind them to consider layout, color, themes, design, messaging, and so on.	
Develop a Solution—Part One • Choose a design • Identify needed materials	• Students settle on the best design; students discuss how it meets constraints and criteria. • Design varies depending on specified end products.	

Develop a Solution—Part Two
• Create a plan; make a sketch
• Build the visual aid or platform
	• A good deal of planning should occur before building this project. Groups need to plan the following.
	  + Overall message
	  + Layout of any visuals
	  + Needed statistics
	  + Storyboard for video or website projects
	  + More specific steps (depending on the actual product)	

Figure 6.4: Where Does All the Water Go? project plan.

continued ▶

Develop a Solution—Part Three
• Present the visual aid or platform
• Get feedback
• Plan modifications to make it better
	• Testing: Get consumer feedback. Depending on the product, this could involve other groups giving feedback via sticky notes on the display or serving as test audiences or users for videos or websites. I strongly suggest groups create mock-ups (simple sketches or diagrams) for this stage rather than a finished product. These allow several modifications before creating the final product.
	• Modification considerations follow.
	   + Students must connect their modifications to consumer feedback information. Create a very simple modification form that asks what the modification is and why it is being done.
	   + Using mock-ups or sketches prior to building can help students modify to create a clearer and more engaging display or platform. This project is one you might expect more modifications than normal. This is acceptable if there is documentation and connection to messaging and constraints and criteria.

Develop a Solution—Part Four • Communicate your results	• This is the overall project goal. Be sure to have some external audience exposure. Other classes can use a display or gallery in a corridor, or a presentation for outside "experts."

Group Size
Four or five students should begin to function effectively in teams; jobs can be similar to those in the Board Game project, but substitute a data engineer with a materials engineer.

Assessment
Follow the general guidelines given in chapter 3's Assessing section (pages 80–91).

Resources
There are many resources for information on the water footprint. These are fairly comprehensive and educator and student friendly. • The Water Footprint Network (https://waterfootprint.org/en/about-us) • Water Footprint Calculator (www.watercalculator.org/education/water-resources-for-educators)

Really Making It Real-World Global Challenges

Students in your classroom live in a highly networked world full of both amazing potential and enormous challenges. They will need to innovate and work together to develop new ideas, products, and ways of doing things. To engineer solutions to the challenges facing them, students will need to understand the people and places both problems and solutions will impact. As Veronica Boix Mansilla of Harvard's Project Zero and Anthony W. Jackson (2014) of the Asia Society note:

> To succeed in this new global age, students will need capacities that include, but go beyond, reading, mathematics, and science; they will need to be far more knowledgeable and curious about world regions and global issues, attuned to diverse perspectives, able to communicate across cultures and in other languages, and disposed to acting toward the common good. Put simply,

> preparing our students to participate fully in today's and tomorrow's world demands that we nurture their global competence, herein defined as the capacity and disposition to understand and act on issues of global significance. (p. 5)

The world is a big place and the challenges we all face in terms of food, security, water, energy, health, climate, and environmental issues are staggering. How do you make all this manageable and authentic for eleven- to eighteen-year-old students? Think *small steps and one project, one place at a time*. Have your students choose a location and an end user. Challenges that center on shared global challenges and their impact and potential solutions in specific places can support both worldview and cultural empathy. Just as importantly, these challenges highlight the value of what students are learning in science and mathematics classes, plus they focus on the potential of engineering to make the world a better place and to help others.

Since 2014, I have worked on developing engineering design challenges that focus on global issues. In 2015, the United Nations and its member countries established a set of seventeen Sustainable Development Goals (SDGs). The SDGs deal with issues of equality, better living standards, and care for the environment and planet, and the United Nations and member countries hope to meet these goals by 2030 to ensure a better future for all. Each goal has multiple targets (169 total) that the United Nations, international organizations, and individual countries monitor using agreed-on indicators. By coupling the global view of the SDGs with a focus on specific places and the need to develop appropriate technology, you can create meaningful projects with manageable solutions.

Think in terms of the following three main ingredients.

1. Global challenges in the United Nations' (n.d.) SDGs
2. Local solutions based on appropriate technology and an understanding of the end user
3. An EDP with a strong focus on the needs of the end user, criteria for technology and solutions, awareness of constraints, and a design approach that emphasizes all aspects of sustainability (environmental, social, and economic)

> Challenges that center on shared global challenges and their impact and potential solutions in specific places can support both worldview and cultural empathy.

In a nutshell, you are moving from problem to solution by following the EDP just as you would in any project. The big shift here is the need to focus on criteria and constraints that the culture and available resources already strongly define. This often leads to an approach that supports sustainability. The need to consider the impact of technology also supports a significant amount of systems thinking, particularly about unintended consequences. These projects are rich in skills and you can develop them to include an unlimited amount of science, social studies, and mathematics content. Before looking at two specific projects, let's look at the three key elements in a bit more detail.

Global Challenges in the United Nations' Sustainable Development Goals

Starting with the United Nations' (n.d.) SDGs conveys to students the idea of a shared global mission and vision. The SDGs are not about the welfare of any one country; the United Nations

> The United Nations' Sustainable Development Goals are a well-designed set of classroom resources and a well-curated compilation of resources that add to any background research.

developed them to ensure a sustainable future for all people. The SDGs focus on issues inherent to social justice, sustainability, and stewardship, and they bring all of the real world into your classroom.

The United Nations' efforts researching, developing, and advocating for the SDGs and their component target areas provide a strong platform of resources for students as they begin their background investigation. From a teacher standpoint, the SDGs provide a well-designed set of classroom resources and a well-curated compilation of resources that add to any background research. You can learn more about the SDGs on the United Nations' SDG website (https://sustainabledevelopment.un.org/sdgs). You can find an extensive collection of SDG-related educator resources on the World's Largest Lesson website (http://worldslargestlesson.globalgoals.org). Even if you opt not to tackle an EDP project with a global focus, I strongly urge you to check out both preceding websites (particularly the latter), and share them with colleagues. Table 6.1 summarizes the SDGs; the preceding websites provide more specifics about targets and issues within each goal.

Table 6.1: The Sustainable Development Goals

SDG	General Description
1—No Poverty	End poverty in all its forms everywhere
2—Zero Hunger	End hunger, achieve food security and improved nutrition, and promote sustainable agriculture
3—Good Health and Well-Being	Ensure healthy lives and promote well-being for all, at all ages
4—Quality Education	Ensure inclusive and equitable quality education and promote lifelong learning opportunities for all
5—Gender Equality	Achieve gender equality and empower all women and girls
6—Clean Water and Sanitation	Ensure availability and sustainable management of water and sanitation for all
7—Affordable and Clean Energy	Ensure access to affordable, reliable, sustainable and modern energy for all
8—Decent Work and Economic Growth	Promote sustained, inclusive, and sustainable economic growth, full and productive employment, and decent work for all
9—Industry, Innovation, and Infrastructure	Build resilient infrastructure, promote inclusive and sustainable industrialization, and foster innovation
10—Reduced Inequalities	Reduce inequality within and among countries
11—Sustainable Cities and Communities	Make cities and human settlements inclusive, safe, resilient, and sustainable

SDG	General Description
12—Responsible Consumption and Production	Ensure sustainable consumption and production patterns
13—Climate Action	Take urgent action to combat climate change and its impacts
14—Life Below Water	Conserve and sustainably use the oceans, seas, and marine resources for sustainable development
15—Life on Land	Protect, restore, and promote sustainable use of terrestrial ecosystems, sustainably manage forests, combat desertification, halt and reverse land degradation, and halt biodiversity loss
16—Peace, Justice, and Strong Institutions	Promote peaceful and inclusive societies for sustainable development, provide access to justice for all and build effective, accountable, and inclusive institutions at all levels
17—Partnerships for the Goals	Strengthen the means of implementation and revitalize the global partnership for sustainable development

Source: United Nations, n.d.

Local Solutions Based on Appropriate Technology and an Understanding of the End User

Knowing where your end users live, understanding how they live, and learning about their resources and values are key to any good design. Technology is never one-size-fits-all. Highly industrialized countries tend to design technologies with a focus on capital, not labor resources. The opposite may be true in many other parts of the world; these countries may be labor-rich and capital-poor. Taking a solution from one place and expecting to work in a very different location and culture neglects the very starting point of good design—*know your end user*. The end user should always be the *subject*, not the object, of your solution.

I describe *Appropriate Technology* (AT) as "small-scale, locally resourced, user-centric technology. It is generally technology designed to support basic human needs such as energy, light, water access, living space and conditions" (Kaiser, n.d.). We know that "every new technology has consequences for society; a paramount goal for AT is for positive consequences to outweigh any unintended negative consequences" (Kaiser, n.d.). Appropriate technology, or technology designed with people as its focus, has been around for a long time. It became a movement due to the work of E. F. Schumacher (1973), author of *Small Is Beautiful*. Schumacher (1973) argued that technology should not be regarded only as a means to an immediate end, but that it must be evaluated in terms of its contribution to a process of production or activity beneficial not only to its immediate users, but also to the society at large. Due to globalization, innovative financing, and the increasing focus on sustainability and social justice, AT has made meaningful strides since the turn of the century. More recent advocates have included Victor Papanek (*Design for the Real World: Human Ecology and Social Change*; 2005) and Amy Smith of the D-Lab program at MIT. The work of Practical Action, an organization Schumacher and colleagues originally named the Intermediate Technology Development Group when they founded it in 1966, realizes much of Schumacher's vision.

Consider the attributes in figure 6.5 to get a better understanding of appropriate technology. I often include an appropriate technology checklist like this one in the student project documents involving local solutions to global challenges. Resources and lesson plans that include appropriate technology follow; you can also visit **go.SolutionTree.com/21stcenturyskills** for live links to these and more resources.

- Practical Action Schools (https://practicalaction.org/schools)
- Engineering for Change (www.engineeringforchange.org)
- The Pachamama Alliance (www.pachamama.org/appropriate-technology)
- Appropedia (www.appropedia.org/Appropriate_technology)

How appropriate is your technology?
Check all those that apply. Ensure your design includes at least five of the following.
- ☐ **Compatible** with local culture
- ☐ **Small scale** so the community should not have to rely on heavy industry or corporate wealth
- ☐ **Local tools and practices** can be used
- ☐ **Local materials and resources are** used as much as possible
- ☐ **Local energy** sources are used
- ☐ **Environmentally friendly**
- ☐ Can be **repaired locally**
- ☐ **Easily maintained**
- ☐ Values **local creativity and design**
- ☐ **Easy to understand and use**
- ☐ **Low cost**
- ☐ Creates **local industry and businesses**
- ☐ **Flexible** for use in many different situations

Figure 6.5: Appropriate technology checklist.

*Visit **go.SolutionTree.com/21stcenturyskills** for a free reproducible version of this figure.*

Disaster-Resilient Housing (Grades 6–12)

The third element in designing this type of project is essentially the same as with other projects throughout this book with a few enhancements. The first is a strong focus on learning about a particular place and culture as part of the Know Your Problem phase. It is up to you whether you want to limit geographic options for these projects. Some teachers like to tie the project in to an area that students are learning about. One teacher I worked with stipulated that students develop solutions to challenges centered around SDG 2 (Zero Hunger) for Spanish-speaking countries since all his students were studying the language. The class learned about a range of Spanish-speaking countries and used some of their language skills to develop logos and labels for their product and, in some cases, to write simple instructions.

Some teachers like to tie the project to an area that students are learning about.

As part of their background research, it often makes sense and is a good opportunity for students to include some statistics and facts (and analysis of these data) about the overall SDG. One other significant factor to keep in mind is that you should challenge students to think about issues and obstacles related to implementing the actual technology. Including this as part of their "Final Design Summary" (pages 222–225), presentation highlights, or both, is a way to get them to consider what is often the biggest challenge in any new design. Implementing appropriate technology can be even more challenging in remote or less-developed parts of the world or in places that function differently than industrialized countries do in terms of infrastructure and support. Asking students to consider these differences brings the project back to understanding the needs and lifestyle of the end user.

Students can do endless projects based on the three key elements: global challenges, local solutions, and the EDP. Figure 6.6 shows a project plan summary that focuses on SDG 11—Sustainable Cities and Communities. Following a similar framework for SDGs projects allows you to identify and focus content connections in your curriculum. SDGs 2, 3, 6, 7, 9, and 11–15 most easily connect to middle and high school science content. All the SDGs are great starting points for engineering design projects that focus on creating data visualizations. You can follow a plan similar to the WFP project (page 159) if you want a strong mathematics connection. Figure 6.7 (page 169) shows a form for researching a location.

Challenge: Disaster-resilient housing	
Primary SDG Focus: **SDG 11**—Sustainable Cities and Communities	**Secondary SDG Focus:** **SDG 13**—Climate Action **SDG 9**—Industry, Innovation, and Infrastructure
Science Content: • Forces and structures • Natural hazards and natural disasters • Global warming • Fluid mechanics (flooding and wind resilience) • Wave mechanics (earthquake resilience) • Plate tectonics (earthquake and tsunami resilience)	**Country Focus:** • Option one—Students choose a location that has endured a natural disaster in the past ten years. They should match their solution to the type of disaster in that region. • Option two—The teacher specifies what type of disaster students design solutions for and groups select a location accordingly. • Option three—The teacher assigns a specific location to each group.
Overall Sequence (Estimate: fifteen classes)	
1. Introduce the SDGs generally.	Give students a broad overview if SDGs are unfamiliar; highlight resources on the United Nations' website (https://sustainabledevelopment.un.org/sdgs) and the World's Largest Lesson (http://worldslargestlesson.globalgoals.org).
2. Give an overview of AT.	See the resources on page 168; I recommend the video "Lucky Iron Fish" (www.youtube.com/watch?v=iY0D-PIcgB4) as an example of AT in general.

Figure 6.6: Disaster-Resilient Housing project plan summary.   continued ▶

3. Introduce issues related to housing and SDG 11.	Visit UN's World's Largest Lesson website (https://bit.ly/2njRYBC). Visit UN's Sustainable Development Goals Knowledge Platform website (https://sustainabledevelopment.un.org/sdg11).
4. Employ a video or TED Talks hook.	Watch "The Warmth and Wisdom of Mud Buildings" (Heringer, 2017). Other options are available online.
5. Do a quick build.	For wind, the Paper Tower of Power activity (page 96) For water, building a house prototype with simple materials, that floats and can hold pennies or stones
6. Research science concepts (background).	See the There Is Always More to Learn section (page 12) in chapter 1 and the Content section (page 62) in chapter 3; have students help identify what they need to know.
7. Choose a country.	Depending on degree of student choice, groups should conduct appropriate research and complete the form in figure 6.7 (page 169). This form is for students in grades 6–9. You may need to add or ask for more detail for students grades 10–12. It still works well as a graphic organizer at all levels.
8. Know your problem. • Clearly define the problem (where the problem exists, the extent of problem, how often the problem arises). • Identify and research end user needs. • Identify constraints, both for the prototype and for an actual solution. • Define criteria.	Much of the end user definition depends on the location. Understanding actual constraints in that location supports a focus on appropriate technology. Remind students that criteria need to reflect the lifestyle, culture, and values of the location.
9. Know your options. • Conduct any additional research about the location, types of storms or disasters, what has been tried in the location and in similar areas. • Brainstorm many options and ideas.	Encourage students to explore what has already been tried. When brainstorming, prompt groups to consider the location's culture and typical housing (that is, materials, colors, architectural features, and so on).

10. Develop a solution. • Sketch the location. • Plan building steps and create the prototype materials list. • Build the housing prototype. • Test the housing prototype. • Modify the housing prototype.	The materials used for prototyping should be chosen to represent (as closely as possible) those available for the actual solution. For instance, straws can stand in for bamboo; foam core can represent wallboard; plastic wrap can stand in for window glass, and so on. Work with students to develop a suitable test for their structure. Test houses designed to float in a sink; pour or spray water on houses designed to resist tropical downpours; create "wind" using a hair dryer, fan, or compressor to test windstorm-resilient houses; make and employ simple shake tables (instructions at https://bit.ly/2JdmQ3S) to test earthquake-resistant buildings. When designing tests, specify the procedure, number of trials, and what is considered successful. If it is impossible to make actual modifications (the building prototype collapsed or sank, for example), students should discuss possible modifications in their final summary.
11. Present the final solution.	Try using groups to present to the rest of the class, who play the role of local government officials or representatives of an international relief organization. Creating a short summary report or marketing brochure is appropriate for high school students. Videos highlighting the benefits of the housing and even showing tests are possible final deliverables.

Visit **go.SolutionTree.com/21stcenturyskills** *for a free reproducible version of this figure.*

About the Location	
Continent	
Country	
Neighbors	
Terrain (coastal, mountains, desert, for example)	
Climate	
Recent Natural Disasters	
Natural Resources	
Extra Notes:	

Figure 6.7: About the country, region, or village form for the Disaster-Resilient Housing project.

continued ▶

About the People	
Average Income	
Typical Jobs	
Education (Completed)	
Life Expectancy	
Health Issues	
Family Life	
Types of Housing	
Information About Culture, Art, and Food	
Extra Notes:	

*Visit **go.SolutionTree.com/21stcenturyskills** for a free reproducible version of this figure.*

As previously mentioned, you can create an extensive number of SDG-related projects. Try to allow some student choice by letting groups choose the location and end user. Table 6.2 may help you get started, but it is far from an exhaustive list. I have seen projects based on most of the SDGs, and all resulted in some amazing student work. You are truly bringing the real world into your classroom with this learning opportunity!

Table 6.2: Additional SDG-Related Project Ideas

SDG	Project Focus
SDG 2— Zero Hunger	• Vertical farming • Nutrition supplement (for example, bar, soup, or juice) • Improved crops • Aquaponics or hydroponics
SDG 6— Clean Water and Sanitation	• Handwashing materials and campaigns • Water filtration • Water access and transport • Improved access to toilet facilities • Water footprint awareness
SDG 7— Affordable and Clean Energy	• Small-scale solar lighting and LEDs • Bicycle-powered devices • Alternative renewable energy devices • Improved cooking techniques
SDG 9— Industry, Innovation, and Infrastructure	• Transportation, including highway and subway development • Assembly-line management • Global supply chain

SDG	Project Focus
SDG 11—Sustainable Cities and Communities	• Lighting for low-income urban or off-grid areas • Disaster-resilient housing • Public transportation • Urban air quality or green cities • Building techniques using local materials • Livable cities
SDG 12—Responsible Consumption and Production	• Upcycling campaigns and ideas • Substitutes for plastics • Life cycle of products analysis • Better manufacturing and production processes (see the How Does Your Production Line Rate? activity, page 175) • Improved packaging design • Circular economy (Ellen MacArthur Foundation at https://bit.ly/2ZZzTwD)
SDG 13—Climate Action	• Finite resources (fossil fuels and increased greenhouse gases) • Carbon emissions • Loss of coastal regions or management and housing redesign
SDG 14—Life Below Water	• Overfishing and fisheries management • Fish farming • Coral bleaching • Ocean acidification • Plastics in oceans
SDG 15—Life on Land	• Desertification or deforestation infographic or video • Biodiversity board project • Upcycled greenhouse from plastic bottles

Engineering Enablement

In this book, I stress that engineering provides a way to solve problems and make our lives better. New technologies—from simple improvements to high-tech robotics—have the potential to help us in many ways and in many places. Engineering enablement is about how we can provide a better quality of life for those who have physical disabilities and for those who manufacture the products that we use every day.

Designing a Prosthetic Hand (Grades 8–12)

The Designing a Prosthetic Hand project (see figure 6.8, page 172) provides a strong focus on empathy by taking the big goal of enabling all to lead happy and productive lives down to a smaller, more personal level. Students are challenged to design a prosthesis for someone who has lost a hand. The prosthetic hand should provide the client one or two of the functions he or she has lost. In its

Project Title: Designing a Prosthetic Hand	Topic: Identifying a specific client or end user, designing a prosthetic hand, and demonstrating its ability to execute one or two functions.
Grade Level: 8–12 Varying levels of difficulty due to specified requirements, prototype testing requirements, and level of understanding of anatomy and physiology	**Estimated Class Time:** Eight to fourteen classes, depending on grade level

Challenge: Develop a prosthetic hand prototype that can do one or two functions most important to the end user.

Note: Students can choose or the teacher can specify the end user. Grade 10–12 students, particularly in anatomy courses, can develop a full case history of a hypothetical client.

Curricular Connections	Skills Focus
Science—levers in the body; forces; anatomy and physiology of the hand **Social Studies**—issues related to disabilities; worldwide incidence of amputation and subsequent support	**Critical Thinking**—identifying key needs; analyzing effective designs; accessible materials and support if device implemented; developing prototype testing procedure **Creativity**—aesthetics of design; possible modifications and superpowers **Spatial Reasoning**—human interaction with designed environments and objects **Collaboration**—working as a team; recognizing common human factors **Communication**—listening to and acting on client feedback **Empathy**—understanding and addressing a basic need to function in society and independence

Overall Plan

Hook	Videos or articles about Hugh Herr (2014), Aimee Mullins (2009), or Alex Truesdell (Adaptive Design, 2018)
Engagement Activity or Quick Build	Students spend a set amount of time at home (or in class) not using their dominant hand. They record and share experiences, challenges, and limitations.
Background Instruction	Providing basic introduction about hand anatomy is important. Basic concepts include bones and joints, role of tendons, muscles that make the fingers move, and the value of an opposable thumb. Get students thinking about contraction and flexion, as these may also be helpful as students design their hand prototype. Your level of instruction depends on the curriculum and grade level. Teaching physics concepts relating to the function of fingers as levers is also important. Focus on concepts of effort force (input), load or resistance force (output), and the fulcrum or pivot point's location. Note: Most limbs in the body function as third-class levers, which have the effort (input) force located between the pivot point and the load. For example, when you lift your arm out straight, the pivot is at your shoulder, the effort is due to the biceps muscle in your upper arm, and the load is your lower arm. Third-class levers allow for speed and range of motion but, unlike first- and second-class levers, there is no multiplicative effect on the input force (mechanical advantage).

Background Research	Additional information on specific hand functions or anatomy concepts; investigation of simple prosthetic hand models in use; challenges in fit and use of prosthetics
Engineering Design Process	
Know Your Problem • Know your end user • Identify constraints • Define criteria	• Get to know (or create) the case history of your client. How did the client lose his or her hand? What does he or she like to do? What does he or she find most challenging? How would he or she like the hand to look? • Specify functions the prosthetic hand must accomplish. • Teacher-supplied materials and costs; end user further defines • Use end user research to develop a list of features related to functions; consider the appearance of the hand and other factors related to ease of use and maintenance.
Know Your Options • Research • Brainstorm	• Conduct additional research to understand specific required functions. • Use a variety of techniques to get students to consider innovative designs; remind students to consider reliability of function, appearance, and any additional (bonus) functions.
Develop a Solution—Part One • Choose a design • Identify needed materials	• Students settle on the best design; students discuss how the design meets constraints and criteria. • Encourage students to keep the design simple (see the following Possible Materials list) and then to troubleshoot and modify.
Develop a Solution—Part Two • Create a plan; make a sketch • Build the prototype	• Develop a sketch of both the overall prosthetic hand plus the following information: how it will attach to the client's arm and the key features that enable function. These are often the two most challenging parts of the design. • Obtain materials and build the hand; check function as you move forward, and make minor modifications on an ongoing basis; encourage students to record these modifications or develop a graphic organizer for documenting the process.

Develop a Solution—Part Three
• Test the hand
• Plan modifications to make the device better
	• Test for function, comfort, and ease of use.
	  + Work with students before they begin building to specify a testing procedure. Specify what they need to do and how many times they must repeat tests. For instance: pick up, hold, and put down a cup or object (grapes are challenging!) five times. Or pick up and throw a ball a certain distance, and so on.
	  + After initial testing, have students choose someone who is not in their group test the hand. The group can help the tester put on the hand, but they cannot provide additional assistance. Note: Attaching the prosthetic hand over actual hand can be challenging; the key thing is not letting your actual hand assist the function of the prosthetic hand.

Figure 6.8: Designing a Prosthetic Hand project plan.

continued ▶

Develop a Solution—Part Three
• Test the hand
• Plan modifications to make the device better
	+ Encourage your students to take notes as they observe how the tester uses the hand and how well it functions. Students should also question the tester about comfort, overall appeal, and ease of use.
	• Modifications
	  + Students must connect their modifications to their testing information (observations or testing data). Create a very simple modification form that asks what the modification is and why it is being done.
	  + Limit modifications; two or three modifications are generally enough.

Develop a Solution—Part Four • Communicate your results	• The best way for students to present their design is for them to describe their client and then demonstrate their hand, highlighting their key goals for function and overall appeal. A brief discussion should include both their modifications and ideas for additions to their design. Limit time and have students consider using a pitch to the audience representing occupational therapists or medical personnel or agencies. • Anatomy teachers may want to request students identify connections between components of a prosthetic hand and an actual hand.

Group Size

Between three and five students works well. Students should begin to function effectively in teams; jobs are a must due to the number of aspects to the project. Suggested jobs are project manager, biomedical engineer, materials engineer, and physical or occupational therapist.

Assessment

Follow the general guidelines in chapter 3's Assessing section (pages 80–91). The following are key considerations specific to this project.
- This is very much a group project and the empathetic connection generally drives responsible involvement by all team members. Each individual's grade generally comes from some content connection work, background research, classroom performance, and final peer assessment.
- Some teachers opt to include a short quiz based on key concepts important in understanding the anatomy and mechanics of the hand. Do not stress detailed vocabulary about anatomy at this point, but you can require it as part of the final presentation for students in grades 10–12.

Materials

• Foam core • Wire, string, twine, yarn • Springs • Velcro • J-hooks, eye hooks • Pins, tacks • Screws • Washers	• Paper clips • Toothpicks • Rubber bands • Large dowels • Sheets of balsa • Small dowels and balsa scraps • Popsicle sticks • Straws and pipe cleaners	• Hot glue and glue guns • Craft glue • Assorted tape • Scissors • Utility or craft knives (optional)

*Visit **go.SolutionTree.com/21stcenturyskills** for a free reproducible version of this figure.*

most extensive version, student groups develop their own case study of a hypothetical patient who has lost a hand.

There are generally a few teacher-given parameters for this version. Students identify the functions that would be meaningful to their client. In less-customized versions, the teacher sometimes requires a function that students must provide to their hypothetical patient or client. Functions both students and teachers have come up with include catching and throwing a ball, opening a door using a doorknob, eating with a fork or a customized utensil, signing his or her name, and even using some sign language.

The two engagement components given in the plan (the video hook and trying to use just one hand) for this project go a long way to support starting from a place rooted in empathy. I strongly recommend doing both. Herr's (2014) TED Talks and other videos never fail to inspire middle and high school students. The work of MacArthur Fellow Alex Truesdell and the nonprofit she founded (Adaptive Design Association) inspire students of all ages. Truesdell and her colleagues (Adaptive Design, 2018) are committed to challenging the notion of disability; they focus instead on the idea that adaptive technologies are not as plentiful, accessible, or customized as they can be. Truesdell's thinking—that it is the technology that needs more engineering—is not that different from Herr's, but her approach is quite different.

Adaptive Design (2018) uses simple materials like cardboard, wood, and fabric to create customized adaptive devices others can replicate around the world or use to replace or modify existing devices as needed. Use any of a number of videos, articles, and resources on the Adaptive Design website (https://adaptivedesign.org) to engage your students in the possibility of simple designs with great impact. Another engagement activity is a homework assignment: ask students to not use their dominant hand for an evening and record some reflections and notes about their frustrations and limitations. You can also do this during class, with either all students participating or with some being potential clients and assistants. Either way, your students will realize we live in a world designed for two hands, and the loss of one can create all sorts of challenges.

I have used this project for students in grades 8–12. The materials are intentionally kept simple to both support an understanding of the physiology and physics involved and to emulate the Adaptive Design (2018) model of accessible, customizable solutions. Some teachers with technology access add a follow-up activity involving 3-D printing of the student-created, simple prototype hands. The curricular connections in this project are the biomechanics and anatomy of the hand, as well as an understanding of forces and simple machines, primarily the lever (the muscular and skeletal structure of the body consists primarily of third-class levers). This project fits well in physical science, anatomy, and physics classes. Again, you can lengthen or shorten this project, but don't skip the empathy component early on. It truly makes this project special. The project plan is in figure 6.8 (page 172).

How Does Your Production Line Rate?—Mathematics (Grades 6–12)

The majority of the objects you use every day are assemblies of various component parts. The pen you write with, the book you are reading, even the candy bar you may have just enjoyed are the products of manufacturing processes that involve putting parts together. Modeling assembly processes provides the basis for projects and activities rich in planning and connections to mathematical concepts of rate, efficiency, and yield.

In addition, the historical, social, and economic impacts of manufacturing and assembly processes make this a great activity for history and social studies classes. The AAAS (2009) points out in *Benchmarks for Scientific Literacy*, "Students should also move from designing and making simple objects to designing and assembling. . . . The importance of planning, coordination, and control should become as evident as the importance of selecting the most appropriate materials and processes." The AAAS (2009) goes on to identify *manufacturing* as an essential feature of human society, and further stresses understanding the reasons for automation depends on being familiar with the existing processes.

The *assembly line* is a production method made popular by Henry Ford; it is still in use in automobile manufacturing and appears in a number of other industrial settings, including electronics manufacturing. Companies use assembly lines when they need to produce large quantities of the same items. You can share a fun example of what can go wrong on an assembly line by having students watch the video about the chocolate factory from the *I Love Lucy* job-switching episode (Oppenheimer, Davis, Carroll, & Asher, 1952).

How Does Your Production Line Rate? is a shorter engineering design activity you can turn into a longer project by adding some requirements and more complex assemblies. During this activity (or project), the teacher gives students a variety of materials to use to create a widget, toy, or device. Students must then deconstruct their creation to develop a series of assembly steps. The challenge in developing the process and setting up the assembly line is that there are fewer workers (students) than parts, requiring some planning for subassemblies. *Subassemblies* consist of groups of parts workers assemble in advance of being included in the larger product. For instance, workers assemble the audio components in your car separately and then add the components to the dashboard as part of the automobile assembly line process. To avoid *bottlenecks* (assembly operations slower than others), students will need to plan both subassemblies and assembly steps carefully. Bottlenecks effectively establish the maximum rate at which the production line can run, so they slow down the entire process. Students will be able to identify bottlenecks pretty quickly when they run trials of their assembly lines because the parts pile up at these points and workers further down the line will be left idle waiting for pieces to reach them.

This activity has many variations, some of which turn it into a longer-term project. Those variations follow.

- Conduct trials to measure the rates of various steps and the overall operation; students then develop equations to model the entire line, with a goal of optimizing production by achieving a high rate with a small likelihood of mistakes.

- Obtain information on *yield* (or the number of good pieces per number of total pieces) and learn about quality issues and ways to maximize overall production by addressing specific causes of defects. These are known as *corrective actions* in manufacturing and production quality reports.

- Use the production line as an analogy to *protein synthesis* in biology classes; this highlights the steps in the protein synthesis process and the speed at which it occurs. It becomes easier to see how errors (mutations) occur.

- Institute a pay-per-piece model to highlight working conditions in many places. This often leads to a discussion of social implications of manufacturing practices, such as the impacts of automation and use of robots to do routine work.

Giving general student instructions for an eight-piece assembly is a good starting point for more complex assemblies and gives students a chance to plan on a smaller scale. Use these instructions if you are looking for a shorter activity that focuses on any of the previously discussed ideas. Visit **go.SolutionTree.com/21stcenturyskills** for the student instructions and other handouts for this project.

Challenge

Use a variety of materials to create something that you then deconstruct and develop assembly steps for, keeping in mind that there are fewer workers (students) than parts, requiring subassemblies.

Group Size

Four students; five if necessary

Time

Between one and five classes based on the variations

Materials

These are the materials.

- Paper clips
- Sticky notes (same color and size)
- Stickers
- Popsicle sticks
- Straws
- Pipe cleaners
- Cotton balls
- Craft foam shapes
- Index cards
- Rocks
- Pieces of pasta
- Uniform small pieces of paper
- Googly eyes
- Beads
- Binder clips
- Buttons
- Uniform small squares of aluminum foil
- Toothpicks
- One roll of tape per group (optional)

Directions

These are the directions.

1. The teacher assembles plastic bags containing fifteen to twenty of the same item.

    Be imaginative about materials. Try to make all of the items in each bag the same color; if you are unable to do this, tell the students color does not matter. Be aware, however, that using different colors often makes it harder for students to build because color often provides helpful cues in the assembly process. Each group needs eight bags of items and more than one group can have bags with the same items.

2. Everyone gets a copy of the instructions and the form in figure 6.9 (page 178). (Visit **go.SolutionTree.com/21stcenturyskills** for instructions to hand out to students.) Explain that you might ask students to give the original they have created and the bags of supplies to another group to see how quickly and correctly they can build them.

Total number produced:
Total number of defective assemblies:
Net assemblies (total number produced minus total number of defective parts):
Production time: ____ minutes Converted to seconds: ____ seconds
Production rate (net parts divided by production time): ____ parts/second
What else do you think it is important to note?
Identify at least two possible process improvements. These are things that will result in fewer mistakes (higher quality) or eliminate bottlenecks (more efficient). Process improvement one: Process improvement two:
Other process improvements:

Figure 6.9: How Does Your Production Line Rate? form.

*Visit **go.SolutionTree.com/21stcenturyskills** for a free reproducible version of this figure.*

3. Students form groups of four factory workers.

4. Each group gets eight plastic bags of supplies from the teacher. Groups open *each* bag and remove one item to use in their creation. They then close all the bags.

   Make sure student groups stop where their handouts indicate.

5. All groups should start assembling at the same time. Groups make a creation with their eight parts and then stop. Once you call *stop*, have the groups note the total number of assemblies produced.

   They get two or three minutes to run their production lines. Two minutes generally works unless the assemblies are complex or the students are not used to planning and working together.

   Since there are eight items in the final assembly and only four workers, some workers will need to do two things, or groups will need subassemblies. Students can lay out all items and design a maximum of two subassemblies beforehand. You can also supply a roll of tape to allow for some taping of parts, but taping is a separate operation (workers can tape more than one spot on the creation at a taping station), so now one other worker must do multiple things.

6. Students remove one item from each bag and build a duplicate creation. They *do not* take apart the first creation.

7. Before beginning the production run, the group should list or map the planned process steps on a separate sheet of paper or on individual note cards at each station.

    An assembled part example should be in view of all members as an exemplar and as a quality control check once production finishes. Groups switch places with another group if the teacher instructs them to.

8. When the teacher says "Start," groups build as fast as they can but try not to make mistakes.

9. When the teacher says "Stop," students count and record the total number of completed creations. If they run out of materials before time is up, they record the time that they stopped assembling.

10. If groups switched lines, have two students from the original group be the inspectors using the original as a guide.

    Two group members stay with the assembly line to serve as production supervisors and answer any questions. (If students did not switch groups, they invite someone from another group to monitor their quality inspection.) Groups record the number of items with defects. The production supervisors subtract the defects from the total items produced to get a *net production* number. They divide this number by the time (in seconds) to get a *rate*. Have students compare rates and discuss answers to the reflection questions.

11. Students fill out the rest of the form, considering possible process improvements.

Reflection Questions

You can debrief students with the following questions.

- Can you identify any bottlenecks in your production line?
- How might you eliminate or improve any bottlenecks?
- Name some bottlenecks you encounter during your daily life.
- Can you imagine performing your part on the assembly line for over eight hours per day?
- What might make your day easier?
- What are two rules that would improve conditions for workers in assembly plants?

In the CLASSROOM

Consider this feedback from Samantha Scutieri, a mathematics and engineering teacher at Union Catholic High School, New Jersey, who centers an entire course based on projects similar to those in this chapter:

> The students are amazed at how much they learn in any given project. When I ask them to reflect at the end, they are surprised at what they came away with and what they actually found enjoyable about the process. Being able to engage students through all learning styles makes them feel appreciated and understood. The students are fully engaged in the projects. They have a stake in them because they helped design them from the bottom, up. The overall projects contain ideas and knowledge from all of the members so everyone is interested. The more hands-on they are, the more engaged everyone is. The students don't even realize how much they are learning about collaboration, communication, and other non-school skills they need to be successful in these unknown careers we are desperately trying to prepare them for (personal communication, August 5, 2018)

Going Forward

This chapter is about doing things with what you know and learning more as you go. That applies to both you and your students. This knowledge provides a way for you to structure projects that center on the EDP and gives students a way to manage the challenge of developing solutions to messy, real-world problems. Your students will learn the curricular content you connect with the project, but they will also learn a lot more.

Take time to reimagine and re-engineer learning in your classroom for a win for both you and your students. As teacher Samantha Scutieri says, "I love that it gives me a chance to learn more about my students. In a class like this, there is more time for discussion, you learn a lot about them and yourself" (personal communication, August 5, 2018).

PART III
REFLECTION

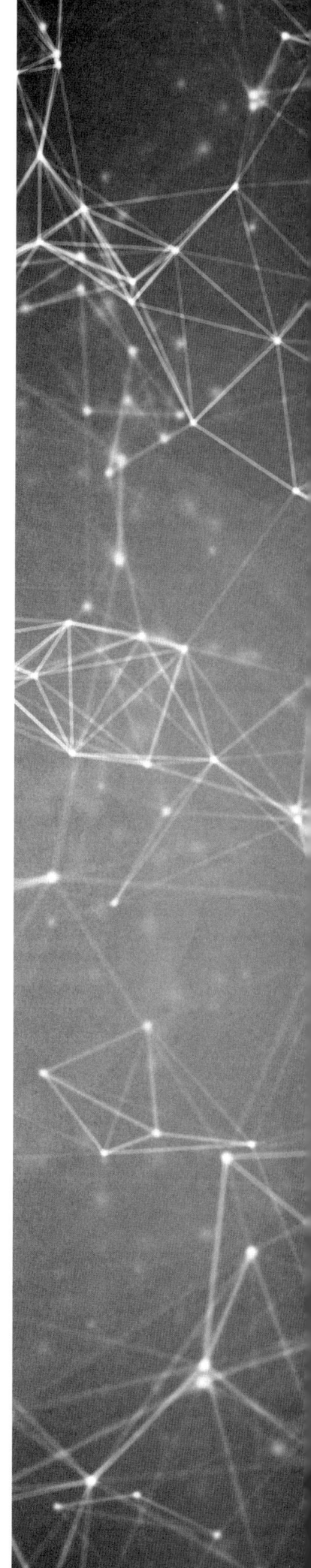

CHAPTER 7
Reflecting On, Revising, and Optimizing Your Curriculum

Success is not final, failure is not fatal: it is the courage to continue that counts.

—Unknown

The final part of the EDP—analyze your test results, modify, and optimize—often gets shortchanged in the classroom. We often shortchange it in our work as educators, too—allowing too little time for reflecting, revising, and optimizing. Promise yourself that you will not skip this phase. It is what will make you see the results realistically and give you the courage to continue your efforts to include engineering design in your classroom. I have been in so many classrooms and have watched all sorts of projects and activities launch. Few first attempts are perfect, but I have never seen one fail to provide some magical learning moments. Don't lose sight of that.

Reflection: The Luxury of Learning

As teachers, we are consummate learners. We learn more about what we teach, who we teach, and how to teach every time we practice our craft. But in the hustle and bustle of the normal school day, the needs of your students, requests from the office, and your life outside the classroom, time for reflection gets lost. I sometimes think finding some time for critical reflection is the biggest challenge you will face when you opt to take this more active, project-based approach. But it is crucial to do so.

> No project, no matter how well it is vetted and designed, will look the same from one classroom to another.

No project, no matter how well vetted and designed, will look the same from one classroom to another. The combination of student choice, creativity, nonlinear thinking, and available resources makes it different every time. This also means some project parts will work well and others, well, not quite so well. I admit to having the luxury to predict which since I see students implement many of the same projects in diverse environments with different teachers. But there is always some new twist—for better or worse.

This is not a full reflection on your overall teaching practice. This involves reflecting about what worked and didn't work in the project; it's more like a project debriefing. Think of this reflection process as focusing on the project components I discussed in chapter 3 (page 59)—challenge, content, and process with the related skills. Add to those the logistics related to implementing the project such as resources, timing, assessment, and classroom management.

Create a checklist that works for you and then be sure you use it. Make it simple, informative, and focused on key aspects of the project. Use a Likert scale (1–5) or an even simpler rating system that leaves room for brief comments. A simple form that works from project to project helps, too. If it is too complicated or too time-consuming, you probably won't use it. The goal here is to get ideas, or even prompts, down on paper as soon as you can during and after the project is over.

To assist you in creating your own reflection checklist, figure 7.1 groups together some key elements to consider. Figure 7.2 (page 186) shows a sample teacher reflection checklist. Include some of these elements in your checklist, but leave room for the unexpected. Use this checklist as a guide if you find it helpful, but remember, your checklist should reflect what you were trying to focus on with your project.

The key to effective reflection is to get your ratings and thoughts all down on paper as soon as possible after completing the project; in some cases, you may even want to complete some of it while the project is underway if you have time and sufficient information. You want to avoid pulling out your project plans next year and having that nagging feeling something was not quite right. The teacher post-project reflection checklist in figure 7.2 is helpful even if you don't go any further than to include it with your plans as a prompt for next year. After all, humans are pretty good at forgetting pain, but we are all pretty miserable when we realize that pain is self-inflicted. If the best you can do is to save yourself from repeating mistakes, your project will still evolve and you will avoid some misery.

> Think of reflection as assessing the journey. Think of the next step or revision as drawing a better map to follow.

The real goal is to learn from your mistakes. Remember the vehicle that contains the skills and content you want to teach I mentioned back in the introduction (page 1)? Don't keep driving it down the same dead end or bumpy road. Reflecting on the path of your project means getting out and looking at where you've been, how far you've come, and making corrections to your route. It means looking at the time, effort, and resources it took to reach the destination, and determining if there might be a more efficient or effective way to get there. We are all fond of saying *it is about the journey*, but in schools, we often determine success based on the final destination or product. Think of reflection as assessing the journey. Think of the next step or revision as drawing a better map to follow. Engage your teacher GPS—you know this landscape. You know the school, the students, and the content. The challenge is to find the best route to success.

	Focus	Key Considerations
Content	Amount	• Teacher given? • Student driven?
	Delivery	• Lecture • Resources • Activities
	Connections	• Were the connections clear? • Were the connections made?
	Understanding	• Was student learning evident? • Are there points that need clarification or amplification?
Process and Skills	Design Challenge	• Did you clearly state the challenge and did students understand it? • Were there too many or too few constraints? • Were students able to define and meet the criteria? • Was the planning manageable for students?
	Creativity and Divergent Thinking	• Was the brainstorming enough and evident? • Did the brainstorming leave room for original thought? • Was the value of brainstorming evident in the project's other phases?
	Building Prototypes	• Was it a collaborative effort? • Did everyone have enough to be busy? • Was testing effective? • Does the testing procedure work? • Did you allow modifying? Were the modifications effective?
	Final Deliverables	• Did the prototypes turn out as expected? • Was the build too difficult? • Was the process documentation enough? Too much? • Was the presentation format effective?
Logistics	Time	• Did you allot enough time for the overall project? • Did the specific steps take longer (or less time) than expected?
	Groups	• Were groups the right size? • Did groups have the right composition? • Were the jobs the correct ones for the project? • Did the group have any management issues?
	Materials	• Consider the consumable materials. Did you need more of anything? What went unused? • Did groups plan and use reusable supplies (tools, scissors, and so on) correctly?
	Assessment	• Did you conduct sufficient formative assessments? Summative assessments? • Was the allocation of group and individual assessments appropriate? • Were the assessments manageable? • Did the assessments reflect student learning?

Figure 7.1: Project elements for teacher reflection.

*Visit **go.SolutionTree.com/21stcenturyskills** for a free reproducible version of this figure.*

Project:		
Content	**Rating**	**Possible Modifications**
Key ideas were clearly stated and understood.	1 2 3 4 5	
The amount of direct instruction was appropriate.	1 2 3 4 5	
Students' hands-on investigation was helpful.	1 2 3 4 5	
Online background resources were sufficient.	1 2 3 4 5	
Content was sufficient.	1 2 3 4 5	
Process and Skills		**Possible Modifications**
The design challenge was clear with the appropriate number of constraints and criteria.	1 2 3 4 5	
Creativity, divergent thinking, and brainstorming were evident.	1 2 3 4 5	
Planning and following the process were evident.	1 2 3 4 5	
There was a collaborative effort, and jobs supported collaboration.	1 2 3 4 5	
The testing procedure was effective.	1 2 3 4 5	
The final prototype met quality expectations.	1 2 3 4 5	
The final pitch presentation reflected on the overall process.	1 2 3 4 5	
Logistics		**Possible Modifications**
The time allotted was sufficient.	1 2 3 4 5	
Was any one part over- or under-planned?	1 2 3 4 5	
The group makeup and jobs were appropriate.	1 2 3 4 5	
The proper materials were available for prototyping.	1 2 3 4 5	
Was there sufficient evidence to make formative assessments?	1 2 3 4 5	
Did the summative assessment reflect student learning?	1 2 3 4 5	
Were assessment components for individuals and groups well designed?	1 2 3 4 5	

Figure 7.2: Sample teacher post-project reflection checklist.

*Visit **go.SolutionTree.com/21stcenturyskills** for a free reproducible version of this figure.*

Revision: If It's Broke, Fix It

In general I am pretty optimistic, but I would be hard-pressed to say you'll hit a home run the first time you're at bat. However, if you practice what you have been teaching, you will hopefully realize failure just means *not yet*. Don't throw out the whole project; modify it with an aim to optimize it. That's why a reflective checklist is so critical. It would be unusual for any of the projects, activities, or suggestions in this book to be a total disaster. I have seen them all implemented, and I would not have included them if I thought they were ineffective. But remember, you just put something that only existed in print and in your head into a highly dynamic environment—one that changes daily, depending on which students are tired, hungry, stressed, or distracted. No one should promise you 100 percent success. You will need to revise to optimize. But you don't have to fix everything all at once. Identify the key issues and make some notes about how you will address them.

> You just put something that only existed in print and in your head into a highly dynamic environment—one that changes daily, depending on which students are tired, hungry, stressed, or distracted.

Let's go back to the vehicle and my advice in the introduction. Think about the main premise of backward design as well. Know where you want to go. Know the project's learning goals. You must keep a destination in mind if your revisions are going to have any real meaning, and it is just as important that you have a place to start. Just as your students need to define a design space using constraints and criteria, you need to consider where things start using your students' prior knowledge and ability to function in different learning environments. These are the constraints you started with, and the criteria for success depend on what learning you hope to see as a result of the project or activity.

Failing to define a starting point and a destination means you and your car full of good ideas will wander aimlessly; no revised route will help. No extra time, additional resources, or perfect students will get you where you want to go if you have no idea where your destination is.

Ask yourself the following questions before tackling your checklist.

- "What content did I want to connect to the project—big ideas, essential questions?"
- "Does the design challenge support the connection?"
- "How much should I expect my students to know about this topic? Will they have any prior correct or incorrect understandings?"
- "Are the level and number of skills I am trying to focus on appropriate for the students in my class?"

Most of the first attempts at engineering design projects I help teachers to troubleshoot have some lack of definition in one of these areas. This lack of definition is unavoidable in many cases because you may need the project to go live to see any gaps. But remember, learning from those mistakes will result in a more robust project next time. Lots of things can go wrong, but there are probably a lot more things that go right. Don't get discouraged! Revising the project will truly make it yours. It is not unlike finally getting the diversity we hope to see in student projects as they develop to fit the group's criteria.

Figure 7.3 lists a few things that could go wrong and some possible revisions I have seen work. These revisions are intentionally general, but you will recognize where they fit with your projects. Tailor revisions to your specific project, but the suggestions in figure 7.3 are a good place to start.

Issue	Possible Revision
Time	
The project took a lot more time than I expected.	1. Always plan for 20 percent more time than you think you need, even if you are experienced. 2. Do not give in (within reason) to requests to adjust deadlines for key tasks. 3. Reassess the overall project for repetitive or extraneous steps. You do not have to equally emphasize every step of the EDP in every project.
A specific phase took far longer than planned.	This usually occurs during the building phase. Try one of the following. 1. Require a detailed plan of the steps before building. Remind students that their planned approach needs to reflect time as a constraint. 2. Institute a limited or no modification rule before testing; constant tweaking negates the impact of planning and delays completion.
Groups	
The project did not involve all students.	1. Reassess group size based on the prevalence of the problem. 2. Assign jobs and responsibilities. This is critical.
The groups were not functioning and needed constant supervision.	1. See numbers 1 and 2 in the preceding cell. 2. Require groups periodically complete the How Are We Doing? form in figure 3.5 (page 73). 3. Institute and remind groups of peer assessment. Consider having students conduct peer assessment midway as well as at the end of the project.
Instruction	
The amount of direct instruction was too low or too high.	1. This is always a challenge; follow the general guidelines given in the section There is Always More to Learn (page 12) in chapter 1 and the section Content (page 62) in chapter 3. 2. Look for gaps or similar issues that came up in students' questions during the project. Try to proactively notice this for future adjustments.
Students had trouble gathering and making sense of the information they needed.	1. Re-evaluate the resources you gave students. 2. Does the framework you provided directly support connecting new ideas? 3. Is there a clear connection to the project? Do students have a need to know?

Assessment	
It was difficult to objectively grade.	1. Identify key engineering notebook forms you will use to evaluate, and consider using or providing exemplars. 2. Re-evaluate your rubrics and do not hesitate to modify them for each project.
I need more evidence that students understand content.	1. Consider formative and summative milestone quizzes throughout the project. 2. Make part of the individual assessment a brief discussion of how key concepts link to the project and prototype. 3. Consider pre- and post-concept (diagnostic) tests.
Materials	
Materials were too costly.	1. Available materials are a constraint; point out what they are and their limitations. 2. Consider requiring a budget constraint and putting a price limit on all materials. This will eliminate waste and poor planning—there is no scrap value!
There were safety or cleaning issues.	1. The biggest issues usually involve cutting tools, so revise your materials list to just scissors if needed. 2. Require any paint and hot glue use in designated places only; keep old plywood boards available to use as surfaces when using these materials.
Process	
There was no evidence of brainstorming.	1. At first, you should intentionally moderate brainstorming sessions and require documentation. 2. Have students participate in some divergent thinking and creativity activities at random and during non-project times. See chapter 4 (page 95) for ideas. 3. Require students to make connections to brainstorming ideas in the "Final Design Summary" reproducible (page 222).
Students made too many changes or did not provide evidence of a planned design.	1. Limit modifications—three maximum— up front. 2. Require students to complete a "Design Modification Request" (page 221) for each change, with the reason for and result of the change.
Not all students presented the final product.	1. Make it clear in job descriptions what everyone will present. 2. Always include time to question students and address each individual student.

Figure 7.3: Possible project revisions.

*Visit **go.SolutionTree.com/21stcenturyskills** for a free reproducible version of this figure.*

Optimization: Re-Engineering to Move Toward Sustainable Change in Your Classroom

You will know when engineering design has become part of your practice; you will constantly see things that can make projects better. You will want to continually re-engineer. So many teachers I have worked with have, at some point, noticed they were pretty much using the EDP to develop and improve projects. I have been in meetings with business development and marketing types where I have been asked to describe what I do and my company's (ProjectEngin LLC) overall mission. When I summarize the EDP, their reaction is invariably that it is similar to how they think. It is how we all think at times, particularly when we have a messy problem to cope with.

Remember those messy problems from chapter 3 (page 59)? A wide range of factors, systems, and human behaviors impact these problems. That sounds like any classroom I have ever been in, including my own. Messy problems involve expected and unexpected interactions. Twenty or so students—need I say more? And messy problems have multiple solutions. That is why you can always modify and adapt, and why so much education research aims at exploring the problem, identifying the options, and trying different solutions.

Do what works for you and your students. As you work to re-engineer how you teach (by introducing engineering design practices and thinking into your classroom), do the following.

- Start the same way you ask your students to begin a project or activity: know your end user—your students

- Keep your students in mind when you design and modify projects.

- Observation is a powerful tool. Use what you see happening with shorter activities as you develop longer-term projects. Look for signs that students want to keep going and be alert to the times when they seem to lack focus.

- Think low walls and high ceilings. Make the project space comfortable for students to jump into but don't limit how high they can go. Don't overload them with details, but remember to give them a good framework to launch from.

- Don't try to save the world in your first project. Low walls and high ceilings apply to you as well. Keep the walls low by starting with a topic you have some expertise in.

- Consider yourself an end user as well. A solution won't work well if it doesn't fit your needs. This is exactly why I never provide fully scripted plans to the teachers I work with. If you want to sustain change in your classroom, you must engineer it to fit your needs.

There are countless options for projects, but keep the EDP in mind and make sure that what you decide to include fits your classroom's constraints (including time, resources, and student ability). Consider the criteria that will define success—learning both skills and content. Be sure your project design starts at the right place and evaluate alternative ideas if it doesn't. Remember, planning a route to your destination is impossible if you have not determined where you are starting.

As a project goes live, be alert to feedback and take time every day to observe students. This is where you are essentially testing your "prototype" and where ideas for improvement begin. Don't obsess over what doesn't work. Missteps just mean it could be better. Moving toward optimization means identifying an issue and making plans for modification, either the next day or the next time you try the project.

Engineers know that they are never done; you can improve every solution. You will no doubt feel the same way as you work through different projects—you can always improve them. But resist the urge to modify too much. I have seen extensive modification result in projects that take too long, lose sight of the learning goals, and become too scripted to allow student choice. What happens is that you will move further away from your project design plan if you modify too much and don't connect it to your constraints, criteria, end user, and all your testing results (including feedback, observations, and student learning). It is not any different than allowing your students to skip the design process, scrap their plans, and try out different things until something seems to work somehow. Tinkering might be helpful in quick builds or in some shorter classroom activities, but it should not be the method you use to design learning experiences.

Think about it—you have essentially designed a prototype, put it out there for testing, and modified it based on the results. You have just re-engineered the *what* and *how* of your teaching. But as teachers, we are always trying to help students learn. Answering the question *Does it work?* has always been tricky. It is why assessment is almost always the elephant in the room when teachers talk about what they are doing and trying.

I am convinced the EDP works, but you need to convince yourself. You can use a test or quiz similar to one you already use to assess students' content understanding after they have learned the material in the context of an engineering design project. Try doing pre- and post-activity assessments of understanding to determine improvement. Listen to your students' questions and discussions as they develop a solution and during the final presentation of their ideas. You and students will have that elusive *aha* moment.

But don't forget that you have also added critical skills to the portfolio of student learning. Their ability to solve problems, work as a team, create something new, and speak about using the process with a sense of ownership and empathy will go far beyond your classroom. Don't just assess; watch students and ask them about the process. You will hear and see things that will convince you this is a powerful learning experience.

As you work to engineer better learning experiences, you will hear some of the following sentiments, similar to those in chapter 1 (page 7), become part of your classroom.

- "It is OK to make mistakes."
- "There are multiple ways to do this."
- "We can make it better."
- "That could create more problems."

These statements indicate that your students are moving beyond the convergent thinking prevalent in many classrooms. They are moving beyond the model of one right answer they are accustomed to and looking for connections, impacts, and options. They are modeling real-world thinking.

In the CLASSROOM

Here is some evidence directly from two teachers and two students about how this powerful learning experience is engaging them.

A high school senior in an engineering design course that follows the guidelines in this book reflected on his experience in an Ugly Christmas Light-Up Sweatshirt project:

> Working in groups was essential to the outcome of our project. Each person in my group was in charge of different tasks. We have a fabric engineer, programmer, design engineer, and project manager. Each job was essential as it contributed greatly to the project. We also learned from other groups by seeing how they programmed their lights or what fabrics they used. If our programmer was confused, he could go to a programmer from a different group to see if they could solve the problem together. —High school student

Mike Martel teaches a curriculum developed to include periodic engineering design projects:

> My students have commented during end-of-the-year reflection discussions that having a project for each unit of study is fun and helpful because of the hands-on nature of project-based learning. They get to use a variety of materials and tools which is extremely engaging, and applying new concepts to solve a problem helps them understand scientific concepts. —Michael Martel, sixth- and seventh-grade science teacher, De La Salle Middle School, Rhode Island (personal communication, August 4, 2018)

Anthony Marmora's classes include group engineering design projects:

> Teaching this way allows students to use their imagination to build things, communicate their ideas with others, and helps them to manage their time in setting and meeting group deadlines. —Anthony Marmora, high school chemistry and engineering teacher, Hudson Catholic High School, New Jersey (personal communication, August 12, 2018)

This student's comment illustrates why letting students know that we can engineer solutions to global issues is so important and empowering:

> My favorite part of this experience was you and your group members are building a product that can solve a problem that is destroying the world. I also like that the ideas for solutions are endless. —Fifth-grade student

Engage your students in learning by letting their need to know guide the process, not by delivering a packaged curricular product to them. Designing your projects to allow student choice (by asking groups to identify different end users or criteria) will create a differentiated learning experience for all your students.

In the CLASSROOM

In speaking about how the use of engineering design projects has impacted his middle school science class at the Academy of Saint Mary, in New Jersey, Robert Zaccone notes:

> It allows hands-on experience that leads to a better understanding of the scientific concepts being explored. It allows for independent thinking, creativity, and uplifts student engagement in the classroom. They also take ownership and responsibility for their own learning. Excitement can be felt throughout the classroom. Student engagement is one of the most important components in education. So, the goal is to hook them to engage; continue to keep them engaged using material important to them. The application of the engineering design process touches on Howard Gardner's theories of different intelligence in how we as humans learn. Whether tactile, visual, audio, etc. . . . You can find ways to engage all different learning styles present in the classroom. (personal communication, October 11, 2018)

Going Forward

Hopefully, your attempts to create new learning experiences have been a success. Or, maybe things didn't go quite right. Remember that neither means the end. Embrace the mindset that says you are never really done and things can always be improved. Borrow ideas from your successes and learn from your failures. Good teaching is highly dynamic, and you are undoubtedly used to all sorts of outcomes. This experience may be different from those you have had with a different topic or with a different approach. There is one big factor that changes from year to year in your classroom, and that is the students. Have the courage to continue learning with them.

Epilogue

We cannot always build the future for our youth, but we can build our youth for the future.
—Franklin D. Roosevelt

Bring the real world into your classroom. It needs to be there because the students you are teaching today are our future. They will be faced with developing solutions to a wide range of problems and designing technologies to improve our lives and safeguard the planet. They will have to make decisions about products, processes, and situations we cannot begin to imagine. They will need to be innovative, entrepreneurial, and able to adapt so they can be productive, happy members of society. In his book *The Innovator's Mindset,* George Couros (2015) tells us:

> In a world where new challenges constantly arise, students must be taught to think critically about what they are facing. They must learn to collaborate with others from around the world to develop solutions for problems. Even more importantly, our students must learn how to ask the right questions. (p. 5)

As a teacher, you can take steps to close the gap between your classroom and the world outside. Connect concepts, embed skills, and model the process of creative problem-solving with engineering design. Centuries ago, students went to school to *find out.* They learned facts and figures and were exposed to scholarly works. As teachers, we need to think about re-engineering our goals; we need to think of our role as enabling students to *figure out* what they need to know.

In the CLASSROOM

As Academy of Our Lady in New Jersey seventh- and eighth-grade science teacher Doris Treuer points out, when using engineering design projects:

> I am teaching students to literally solve problems in an ever-changing world. Students are learning to work collaboratively, take risks, make mistakes, develop goals, and work within existing constraints. Students are not silently waiting for information to be poured into their brains, they are now actively engaged in the education process. They are researching and grappling with real-world problems. They are inventing, creating, designing, building, and enjoying the process. (personal communication, September 30, 2018)

As teachers, we need to provide a framework for knowledge and a process for students to find out more as they learn to figure things out. It is a big shift, but hopefully this book has given you some ideas to move forward and navigate a path from where you are now to where you want your students to be. I promise you will be inspired to do more of it!

In *Future Wise*, author David N. Perkins (2014) takes an upbeat view of this challenge:

> Educating for the unknown, far from an unapproachable paradox, can be an alluring and inspiring agenda. Rather than counseling despair, educating for the unknown favors a vision of learning aggressive in its effort to foster curiosity, enlightenment, empowerment, and responsibility in a complex and dynamic world. It favors a broad and visionary reach for meaningful learning. (p. 23)

Your students can now *find out* from countless sources. Your goal is to support the background, curiosity, engagement, and empowerment students need to *figure out* what they can do with all that information.

For many of us who spend our lives working with young people, it often just takes that one student to let us know we have made a difference to renew our commitment to trying new things in our efforts to reach all students. By modeling the idea that problems are best solved by teams of people with different talents and backgrounds, you will find you have the opportunity to make that difference. It has happened to me; one of the most rewarding parts of my work is the number of times teachers tell me they have seen it happen in their classrooms.

> You don't teach a class; you teach young people.

You don't teach a class; you teach young people. Making a difference starts with how you impact them. It starts with providing engaging and empowering projects that give them the confidence and skills to design a better future.

In the CLASSROOM

This description from Joanne Cavera, a high school biology teacher who has developed her own engineering design course at St. Joseph Regional High School in New Jersey, is what makes change worth the effort:

> Student B came to me at the beginning of the year last year, no eye contact, hiding behind his hair, mumbled that he is bad at mathematics and never did well in science and was afraid he would fail the engineering class as he was always a mediocre student. He was very uncomfortable working in a group and struggled to communicate with his fellow students. But he wanted to take the class because he liked to think about different ways of doing things. As the year progressed he became more and more confident. By the end of the year he was volunteering to be the project manager, stayed in during lunch time, and spent extra time after school to work on projects. The last day of class he stayed after to talk to me. He looked me in the eye and thanked me for the class. He found out that he could work well on a team, was a good and organized leader and, for the first time, he achieved success in a high school class, both academically and socially. The difference in him was incredible. This is what makes it all worth it. (personal communication, October 11, 2018)

APPENDIX A
Action Plan Summary

Designing the Future—Action Plan

Here is a summary of steps that can help you incorporate the ideas in this book into your classroom practices and culture, and then help you move forward to include projects.

	Step 1	Priority	Notes
Deciding *why*: Creating a vision	Real-world connections		
	Transdisciplinary learning		
	Highlight 21st century skills		
	Project-based learning		
	Student choice and differentiation		

Step 2		Possible Activities
Deciding *how*: Focusing on classroom and culture	Learning from failure	• Paper Tower of Power • Three-Legged Stool • Flight of the Table Tennis Ball
	Self-directed learning; developing need to know; observing others	Use concept maps; introduce a broad topic and allow different groups to find different information.
	Multiple solutions	Find three ways to get from classroom to cafeteria; investigate pros and cons.
	Systems thinking; considering impacts	• "Systems Thinking: A Cautionary Tale (Cats in Borneo)" video (Sustainability Illustrated, 2014) • Resources from the Waters Foundation (2017) and BrainPOP Institute of Play (https://educators.brainpop.com/tag/institute-of-play)
	Always thinking of improvement	Make the things around us (desks, lunch line, homework) better

Step 3		Possible Activities
Introducing the EDP: Step one—Know your problem.	Defining what matters in a problem	Five Ws
	Knowing your end user	Name Your Pain
	Understanding constraints and criteria	• Constraints that lead to objects in room • Reverse design space • Criteria ranking
Step two—Know your options.	Creativity and divergent thinking activities and projects	• Stars and Stripes (and Dots) • Chindogu • SCAMPER This!
Step three—Develop a solution.	Prototyping—making your ideas visible	• Ready, Set, Design • Sketches • Cardboard Arcade
	Testing and modifying	• ModiFly
	Communicating and collaborating	• LEGO Person Challenge • No Words

Step 4		Project Planning
Engineering solutions: Project time (use forms indicated)	Overview—"Project-Planning Canvas" reproducible (page 62); describe challenge and final product	• Content • Skills • Process
	Details—"Project-Planning Template" reproducible (page 206)	**Notes**
	Hook	
	Quick build	
	Direct instruction	
	Background research	
	Engineering notebook forms (in appendix C, page 213)	
	Group size	
	Jobs	
	Materials	
	Assessment	

REPRODUCIBLE

Step 5 (Post-project reflection checklist, page 186)			Notes
Reflect, revise, and re-engineer: Modify to optimize	Reflect	Content	
		Process and skills	
		Logistics	
	Revise	Figure 7.3 (page 188–189)	
	Re-engineer	Modification one	
		Modification two	
		Modification three	

Designing the Future © 2020 Ann Kaiser • SolutionTree.com
Visit **go.SolutionTree.com/21stcenturyskills** to download this free reproducible.

APPENDIX B
Project Planning

This template is the starting point for all of the project descriptions and plans found in chapters 5 (page 125) and 6 (page 149). The template helps provide an overview of any project you will plan. It is meant to provide a place where you can identify the content and skills you hope to reinforce and the general "flow" of the project from introduction through the actual EDP and your goals for assessment. Under the EDP, I generally tie evidence to EDP forms (or other documentation) and artifacts such as the prototype. Use the Assessment section to highlight whatever skills and content will be assessed individually or by groups. You can also note what parts of the process and product might have some individual assessment (although most will be group assessment components). This form is basically a snapshot of your project.

Project-Planning Template

Project Title:

Topic:

Grade Level:

Estimated Class Time:

Challenge:

Curricular Connections	Skills Focus
Science	Critical Thinking
Mathematics	Collaboration
Art	Creativity
Social Studies	Communication
ELA	Empathy
Other:	Global View
	Systems Thinking
	Other:

Overall Plan	
Hook	
Engagement Activity or Quick Build	
Background Instruction	
Background Research	

Engineering Design Process	
Know Your Problem • Know your end user • Identify constraints • Define criteria	
Know Your Options • Research • Brainstorm	
Develop a Solution—Part One • Choose a design • Identify needed materials	

Develop a Solution—Part Two • Create a plan; make a sketch • Build the prototype	
Develop a Solution—Part Three • Present the prototype • Get feedback • Plan modifications to make it better	
Develop a Solution—Part Four • Communicate your results	

Group Size

Possible Jobs

Materials (attach a list if necessary)

REPRODUCIBLE

Assessment	
Individual	**Group**
Skills	Skills
Content	Content
Process	Process
Product	Product

Resources and Notes

APPENDIX C
Engineering Notebook Forms

Engineering Notebook Checklist

The following checklist represents the kind of documentation that normally appears in an engineering notebook. This is a great starting place, but feel free to add or modify what you feel you need in order to fully document the EDP. The Assigned To column simply means that the student is responsible for making sure the necessary information is there and that it is included in the notebook. Multiple students can work on any one form or document and many are assigned to all. Samples of the forms indicated with an asterisk are included in this appendix. Use the forms included in grades 5–12 with modifications for project complexity and project-specific questions if needed.

Name of Form	Assigned To	Date Completed	Initials
Jobs and Responsibilities			
Background Research			
Ranking Criteria (page 109)			
Brainstorming Summary* (page 215)			
Design Ranking* (page 216)			
Initial Design Plan* (pages 217 and 218)			
Project Task Planner* (page 219) (or timeline)			
Detailed Sketch or Key Feature			
Materials List			
Daily Summary* (page 220)			
Prototype Feedback (page 49)			
Prototype Testing (page 49)			
Design Modification Request* (page 221)			
Final Design Summary* (pages 222–225)			

*Note: The forms with an asterisk are included in appendix C. Visit **go.SolutionTree.com /21stcenturyskills** to access all the forms that appear here with page numbers.*

Designing the Future © 2020 Ann Kaiser • SolutionTree.com
Visit **go.SolutionTree.com/21stcenturyskills** to download this free reproducible.

Brainstorming Summary

Some teachers like to use a form like this for groups to summarize key ideas from their brainstorming session.

Team members
What is the problem you are considering? Answer in one sentence.
What are your notes from the brainstorming? Only one team member records the notes.
After brainstorming, take a picture of the workspace or save your sticky notes. Then answer the following questions. What were three specific issues or aspects of the problem?   What were the four to six design ideas that appeared most often during the brainstorming session?   What are at least four key features of your design? Describe them in words or sketch them.

Designing the Future © 2020 Ann Kaiser • SolutionTree.com
Visit **go.SolutionTree.com/21stcenturyskills** to download this free reproducible.

Design Ranking

After groups have come up with multiple options for a solution, they should sketch and give each a name or number. Then they complete this form, either individually or as a team, using a 0–5 rating to indicate how well a potential design meets each criterion. Note: groups should always rank criteria (Know Your Problem, page 103) before they use this form. The top three criteria from that ranking should be on this form.

0: Does not meet any criteria	3: Meets, or nearly meets, two criteria
1: Meets, or nearly meets, one criterion	4: Nearly meets all three criteria
2: Meets one or two criteria	5: Meets all three criteria

	Model Name:	Model Name:	Model Name:	Model Name:	Model Name:
Criterion:					
Criterion:					
Criterion:					
Vote Total					

Designing the Future © 2020 Ann Kaiser • SolutionTree.com
Visit **go.SolutionTree.com/21stcenturyskills** to download this free reproducible.

Initial Design Plan

Teams should complete this form after they have identified constraints and criteria and settled on their solution, but before doing any building. The Initial Design Plan is their entry ticket into the building phase and provides evidence of planning and connecting back to the design space.

Team members and job titles
Criteria: List the top five things your team must keep in mind to develop a good solution. List these criteria in order from the most important to the least important. Use a Design-Ranking Form to complete this section.
The solution's four most important features: 1.
2.
3.
4.
The solution's least important feature: 5.
Constraints: What limitations and requirements must the solution meet? List no more than six constraints.
Design statement: What will your team do to develop a solution for the problem? You may draw a sketch, or create bullet points or paragraphs to create a clear picture of your plan.

Meeting criteria: How will your proposed solution meet the identified criteria? Specifically name the criteria and the ways they are being addressed. Use the list form to complete this section.

Science concepts: List any science concepts you need to understand and apply to meet this challenge.

1.

2.

Additional background information: Identify any data, statistics, cultural, or historical information you may need to research.

Anticipated problems: What issues or roadblocks will your team have to overcome to succeed? For example, how or where will you get materials? How might a scaled model inaccurately represent the actual product or solution? How do you plan on getting the information or product to the end users?

Project Task Planner

Engineering design projects are composed of multiple steps, so it makes sense to help groups plan. This plan should be flexible, but it does serve as a reminder for making efficient use of time every day. Alternatively, groups can create a timeline or a kanban board.

Project Title:	Company Name:

Team Members

Task (What needs to be done?)	Due Date	Assigned To (Initials)	Date Done	Done By (Initials)

Designing the Future © 2020 Ann Kaiser • SolutionTree.com
Visit **go.SolutionTree.com/21stcenturyskills** to download this free reproducible.

Daily Summary

This form is optional, but many teachers find that it provides a good end to each day and a good beginning to the next session. It also acts as a running log to help you see what progress a group is making. This form contains spaces for entries covering three days.

Directions: Complete this form daily or as the teacher requests. Write one or two sentences per question.
Company name or logo:
Form completed by:　　　　　　　　　　　　　　　　　　　　　　　　　　　**Date:**
What work did you do today?
How did your work help push your project closer to completion?
What work must you do tomorrow?
Form completed by:　　　　　　　　　　　　　　　　　　　　　　　　　　　**Date:**
What work did you do today?
How did your work help push your project closer to completion?
What work must you do tomorrow?
Form completed by:　　　　　　　　　　　　　　　　　　　　　　　　　　　**Date:**
What work did you do today?
How did your work help push your project closer to completion?
What work must you do tomorrow?

Designing the Future © 2020 Ann Kaiser • SolutionTree.com
Visit **go.SolutionTree.com/21stcenturyskills** to download this free reproducible.

Design Modification Request

Students complete this form after initial prototype testing in order to identify and record planned modifications. They should have the teacher's signature before modifying.

Team member authorizing change:
Teacher authorization signature:
Reason for modification (What issue was the product or solution meant to resolve by the modification?)
Modification sketch or description (Highlight the areas being modified. Attach a separate sheet of paper if needed.)

Before modification:	**After modification:**

Retest (Using the same procedure that prompted the modification, retest or re-evaluate the product or solution after making the modification. Record the retest data in the following space.)
Modification results (Describe how the product or solution performed differently as a result of the modification in the following space.)
Circle one to show whether the modification request is accepted or rejected.
Accepted   Rejected

Designing the Future © 2020 Ann Kaiser • SolutionTree.com
Visit **go.SolutionTree.com/21stcenturyskills** to download this free reproducible.

Final Design Summary

This form brings students back to the initial design space and asks them to summarize how their planned prototype evolved into a solution.

Group name:	
Team Member	**Job Title or Responsibilities**

Problem statement (Name your challenge.)

Criteria: On the left, list the top three criteria that your team developed and listed on your Initial Design Plan. Fill in the spaces on the right to show how you met each criterion.

Criterion	How We Met It

Constraints: On the left, list three constraints listed on your Initial Design Plan. Fill in the spaces on the right to show how you met each constraint.

Constraint	How We Met It

Final design product or solution: Give a detailed description or make a detailed drawing. Label drawings to highlight the important details.

Modifications or changes: List any significant changes from your Initial Design Plan. How is your final version better at meeting your design goals?

Change	*Why* or *How* It Is Better

Most significant concepts: List the two or three ideas that you used the most when designing your solution.

References and Resources

Accreditation Board for Engineering and Technology. (2018). *Criteria for accrediting engineering programs.* Accessed at www.abet.org/wp-content/uploads/2018/11/E001-19-20-EAC-Criteria-11-24-18.pdf on May 20, 2019.

Adaptive Design. (2018). *Who we are.* Accessed at www.adaptivedesign.org/whoweare on October 22, 2018.

Aggarwal, P., & O'Brien, C. L. (2008). Social loafing on group projects: Structural antecedents and effect on student satisfaction. *Journal of Marketing Education, 30*(3), 255–264.

Albrecht, J. R., & Karabenick. S. A. (2017). Relevance for learning and motivation in education. *Journal of Experimental Education, 86*(1), 1–10. Accessed at www.tandfonline.com/doi/ref/10.1080/00220973.2017.1380593?scroll=top on March 14, 2019.

American Association for the Advancement of Science. (n.d.). *Welcome to the AAAS project 2061 science assessment website.* Accessed at http://assessment.aaas.org/pages/home on July 6, 2018.

American Association for the Advancement of Science. (1990). *Science for all Americans.* New York: Oxford University Press. Accessed at www.project2061.org/publications/sfaa/online/sfaatoc.htm on July 28, 2019.

American Association for the Advancement of Science. (2009). *Benchmarks for scientific literacy.* Accessed at www.project2061.org/publications/sfaa/online/chap8.htm on May 24, 2018.

American Literature. (n.d.). *The three little pigs.* Accessed at https://americanliterature.com/childrens-stories/the-three-little-pigs on June 2, 2019.

Andrews, R. (1990). *The concise Columbia dictionary of quotations* (3rd ed.). New York: Columbia University Press.

Association for Supervision and Curriculum Development. (2011). *What is performance assessment?* Accessed at https://pdo.ascd.org/lmscourses/PD11OC108/media/Designing_Performance_Assessment_M2_Reading_Assessment.pdf on May 31, 2019.

Aston, D. H., & Long, S. (2011). *A butterfly is patient.* San Francisco: Chronicle Books.

Attari, S. Z. (2014). Perceptions of water use. *Proceedings of the National Academy of Sciences (PNAS), 111*(14), 5129–5134.

Averell, L., & Heathcote, A. (2011). The form of the forgetting curve and the fate of memories. *Journal of Mathematical Psychology, 55*(1), 25–35.

Barron, B., & Darling-Hammond, L. (2008). *Teaching for meaningful learning: A review of research on inquiry-based and cooperative learning.* Accessed at https://backend.edutopia.org/sites/default/files/pdfs/edutopia-teaching-for-meaningful-learning.pdf on September 17, 2018.

Benson, T. A. (n.d.). *Developing a systems thinking capacity in learners of all ages.* Accessed at http://citeseerx.ist.psu.edu/viewdoc/download?doi=10.1.1.535.9175&rep=rep1&type=pdf on March 14, 2019.

Benyus, J. M. (1997). *Biomimicry: Innovation inspired by nature.* New York: Morrow.

Bigelow, K. (2012). Designing for success: Developing engineers who consider universal design principles. *Journal of Postsecondary Education and Disability, 25*(3), 211–225.

Boboltz, S., & Yam, K. (2017). *Why on-screen representation actually matters*. Accessed at www.huffpost.com/entry/why-on-screen-representation-matters_n_58aeae96e4b01406012fe49d on July 2, 2019.

Briggs, S. (2014). *Socratic questioning: 30 thought-provoking questions to ask your students*. Accessed at www.opencolleges.edu.au/informed/features/socratic-questioning on March 13, 2019.

Buck Institute for Education. (n.d.). *Rubrics*. Accessed at www.bie.org/objects/cat/rubrics on October 24, 2018.

Buck Institute for Education. (2013). *Research summary on the benefits of PBL*. Accessed at www.bie.org/object/document/research_summary_on_the_benefits_of_pbl on February 4, 2019.

Bureau of Labor Statistics. (2010). *Persons with a disability: Labor force characteristics—2009*. USDL10-1172 [Press release]. Accessed at www.bls.gov/news.release/archives/disabl_08252010.pdf on March 25, 2019.

Burnett, M. (Writer), & Spirko, C., Fuchs, K., & Carter, A. (Directors). (2016). Scrub Daddy pitch [Television series episode]. In Newbill, C., Brunertt, M., Lingner, Y., & Gurin, P. (Executive Producers), *Shark Tank*. New York: American Broadcasting Company. Accessed at https://youtube.com/watch?v=ggi3yfUv0Mo on October 30, 2018.

Burzstyn, L., Egorov, G., & Jensen, R. (2016). *Cool to be smart or smart to be cool? Understanding peer pressure in education*. Accessed at http://home.uchicago.edu/~bursztyn/Bursztyn_Egorov_Jensen_2016_07.pdf on February 28. 2019.

Busteed, B. (2013). *The school cliff: Student engagement drops with each school year* [Blog post]. Accessed at http://news.gallup.com/opinion/gallup/170525/school-cliff-student-engagement-drops-school-year.aspx on Aril 12, 2018.

Care, E., Kim, H., Anderson, K., & Gustafsson-Wright, E. (2017). *Skills for a changing world: National perspectives and the global movement*. Brookings Institution. Accessed at https://brookings.edu/wp-content/uploads/2017/03/global-20170324-skills-for-a-changing-world.pdf on March 25, 2019.

Carle, E. (1994). *The very hungry caterpillar*. New York: Phiomel.

Carnegie Mellon University Eberly Center. (2016). *Using concept tests*. Accessed at www.cmu.edu/teaching/assessment/assesslearning/concepTests.html on February 4, 2019.

Cheng, B. H., Ructtinger, L., Fujii, R., & Mislevy, R. (2010). *Assessing systems thinking and complexity in science*. Menlo Park, CA: SRI International. Accessed at https://ecd.sri.com/downloads/ECD_TR7_Systems_Thinking_FL.pdf on September 5, 2018.

Chindogu. (n.d.). Accessed at www.chindogu.com on September 12, 2018.

Clear, J. (n.d.). *The weird strategy Dr. Seuss used to create his greatest work*. Accessed at https://jamesclear.com/dr-seuss on August 9, 2018.

Costa, A. L., & Kallick, B. (n.d.). *Describing 16 habits of mind*. Accessed at www.ccsnh.edu/sites/default/files/content/documents/CCSNH%20MLC%20HABITS%20OF%20MIND%20COSTA-KALLICK%20DESCRIPTION%201-8-10.pdf on March 15, 2019.

Continuous Improvement Toolkit. (n.d.). *Pugh matrix*. Accessed at https://citoolkit.com/articles/pugh-matrix on June 1, 2109.

Couros, G. (2015). *The innovator's mindset: Empower learning, unleash talent, and lead a culture of creativity*. San Diego, CA: Burgess.

Darling-Hammond, L. (2017, March). *Developing and measuring higher order skills: Models for state performance assessment systems*. Washington, DC: Council of Chief State School Officers.

Daum, K. (2016). 37 quotes from Thomas Edison that will inspire success. *Inc*. Accessed at https://inc.com/kevin-daum/37-quotes-from-thomas-edison-that-will-bring-out-your-best.html on July 6, 2018.

Desjardins, J. (2019). *How much data is generated each day?* Accessed at www.weforum.org/agenda/2019/04/how-much-data-is-generated-each-day-cf4bddf29f on June 7, 2019.

Dr. Suess. (1960). *Green eggs and ham*. New York: Penguin Random House.

Dr. Seuss. (1975). *Oh, the thinks you can think!* New York: Beginner Books.

Dryden, H. L. (1965) *Theodore von Kármán 1881–1963: A biographical memoir*. Washington, DC: National Academy of Sciences. Accessed at www.nasonline.org/publications/biographical-memoirs/memoir-pdfs/von-karman-theodore.pdf on July 20, 2018.

Eberle, B. (1996). *Scamper: Creative games and activities for imagination development*. Waco, Texas: Prufrock Press.

Educational Research Center of America. (2016). *STEM: Classroom to career—Opportunities to close the gap.* Accessed at https://ngcproject.org/sites/default/files/stem_long_report_final_0.pdf on December 31, 2018.

Eveleth, R. (2015). Group projects and the secretary effect. *The Atlantic.* Accessed at www.theatlantic.com/education/archive/2015/01/group-projects-and-the-secretary-effect/384104 on May 29, 2019.

Fessenden, M. (2015). A tired brain could actually be more creative. *Smithsonian.* Accessed at https://smithsonianmag.com/smart-news/tired-brain-could-actually-be-more-creative-one-180954802 on November 3, 2018.

Fluckiger, J. (2010). Single point rubric: A tool for responsible student self-assessment. *Delta Kappa Gamma Bulletin, 76*(4), 18–25.

Foster, S. (2019). *Women in STEM: Critical to innovation.* Accessed at www.globalpolicyjournal.com/blog/10/01/2019/women-stem-critical-innovation on May 29, 2019.

Franklin Institute. (n.d.). *Edison's lightbulb.* Accessed at https://fi.edu/history-resources/edisons-lightbulb on July 9, 2018.

Friedman, R. (2015). *Your brain's ideal schedule* [Podcast]. Accessed at https://hbr.org/ideacast/2015/03/your-brains-ideal-schedule.html on April 1, 2019.

Frymier, A. B., & Schulman, G. M. (1995). "What's in it for me?" Increasing content relevance to enhance students' motivation. *Communication Education, 44*(1), 40–50.

Gagnier, K., & Fisher, K. (2016, July). *Spatial thinking: A missing building block in STEM education* [White paper]. Johns Hopkins School of Education. Accessed at http://edpolicy.education.jhu.edu/wp-content/uploads/2016/07/SpatialthinkingmastheadFINAL.pdf on September 20, 2018.

Geyer, R., Jambeck, J. R., & Law, K. L. (2017). Production, use, and fate of all plastics ever made. *Science Advances, 3*(7). Accessed at http://advances.sciencemag.org/content/advances/3/7/e1700782.full.pdf on October 1, 2018.

Giertz, S. (2018). *Simone Giertz: Why you should make useless things* [Video file]. Accessed at https://ted.com/talks/simone_giertz_why_you_should_make_useless_things on September 4, 2018.

Gnesdilow, D., Evenstone, A., Rutledge, J., Sullivan, S., & Puntambekar, S. (2013). *Group work in the science classroom: How gender composition may affect individual performance.* Accessed at http://ildl.wceruw.org/publications/Gnesdilow_etal_2013_CSCL.pdf on March 24, 2019.

Gobble, T., Onuscheck, M., Reibel, A. R., & Twadell, E. (2016). *Proficiency-based assessment: Process, not product.* Bloomington, IN: Solution Tree Press.

Godin, S. (2014). *Connecting dots (or collecting dots)* [Blog post]. Accessed at https://seths.blog/2014/04/connecting-dots-or-collecting-dots on July 9, 2018.

Goldberg, D. E., & Somerville, M. (2014). *A whole new engineer: The coming revolution in engineering education.* Douglas, MI: ThreeJoy Associates.

Golding, C. (2009). *Integrating the disciplines: Successful interdisciplinary subjects.* Melbourne, Australia: Centre for the Study of Higher Education, University of Melbourne.

Grace Communications Foundation. (2017). *Water resources for educators.* Accessed at www.watercalculator.org/education/water-resources-for-educators on October 18, 2018.

Grant, A. (2016) *Originals: How non-conformists move the world.* New York: Penguin Books.

Greim, R. (2010). *MIT Toy Lab 2010 product design class* [Video file]. Accessed at https://vimeo.com/11776105 on October 27, 2018.

Grimm, J. L., & Grimm, W. C. (2013). *Kinder- und hausmärchen (1812–1815).* CreateSpace.

Hammer, B. (2019). *5 tips for helping students of all ages find credible online sources* [Blog post]. Accessed at https://blog.edmentum.com/5-tips-helping-students-all-ages-find-credible-online-sources on May 22, 2019.

Hancock, D. (2004). Tame problems and wicked messes: Choosing between management and leadership solutions. *RMA Journal.* Accessed at https://slideshare.net/Hank5559/tame-problems-wicked-messes-choosing-between-management-and-leadership-solutions-2577646 on November 1, 2018.

Helen Arkell Dyslexia Charity. (n.d.). *Famous dyslexics.* Accessed at www.helenarkell.org.uk/about-dyslexia/famous-dyslexics.php on August 20, 2019.

Hendley, V. (1998). The importance of failure. *American Society for Engineering Education Prism, 18,* 18–23.

Heringer, A. (2017). *Anna Heringer: The warmth and wisdom of mud buildings* [Video file]. Accessed at https://ted.com/talks/anna_heringer_the_warmth_and_wisdom_of_mud_buildings/up-next?language=en on April 4, 2019.

Herr, H. (2014). *Hugh Herr: The new bionics that let us run, climb and dance* [Video file]. Accessed at https://ted.com/talks/hugh_herr_the_new_bionics_that_let_us_run_climb_and_dance on October 18, 2018.

History.com editors. (2009). Wright brothers. *History*. Accessed at https://history.com/topics/inventions/wright-brothers on July 9, 2018.

Hodges, L. C. (2017). *IDEA paper #65: Ten research-based steps for effective group work*. Accessed at https://ideaedu.org/Portals/0/Uploads/Documents/IDEA%20Papers/IDEA%20Papers/PaperIDEA_65.pdf on January 7, 2019.

Hoekstra, A., Chapagain, A. K., Aldaya, M. M., & Mekonnen, M. M. (2011). *The water footprint assessment manual: Setting the global standard*. London: Earthscan.

Hoekstra, A., & van Heek, M. (2017). *Product gallery*. Accessed at https://waterfootprint.org/en/resources/interactive-tools/product-gallery on October 17, 2018.

Holm, M. (2011). Project-based instruction: A review of the literature on effectiveness in prekindergarten through 12th grade classrooms. *InSight: Rivier Academic Journal, 7*(2). Accessed at www2.rivier.edu/journal/ROAJ-Fall-2011/J575-Project-Based-Instruction-Holm.pdf on April 20, 2018.

Imagination. (n.d.). *Kick off your global cardboard challenge*. Accessed at https://cardboardchallenge.com on September 13, 2018.

The Information Blanket. (n.d.). Accessed at www.informationblanket.com on July 1, 2019.

Institution of Civil Engineers. (2018). *What is civil engineering?* Accessed at https:// ice.org.uk/what-is-civil-engineering on September 17, 2018.

Johnson, D. W., Johnson, R. T., & Smith, K. A. (1998). Cooperative learning returns to college: What evidence is there that it works? *Change, 30*(4), 26–35.

Jones, S., Weissbourd, R., Bouffard, S., Kahn, J., & Ross, T. (2018). *For educators: How to build empathy and strengthen your school community*. Cambridge, MA: Harvard Graduate School of Education. Accessed at https://mcc.gse.harvard.edu/files/gse-mcc/files/empathy_brief_for_schools2.pdf?m=1448053946 on September 7, 2018.

Kaiser, A. D. (n.d.). *What is Appropriate Technology?* Accessed at www.projectengin.com/Site/pdf/Teacher%20Toolbox/2%20What%20is%20Appropriate%20Technology.pdf on July 1, 2019.

Kaiser, A. D. (2014). *A modular approach to using the engineering design process in secondary science curriculum: Experiences in Singapore and the United States*. Paper presented at the 2014 Institute of Electrical and Electronics Engineers Frontiers in Education (IEEE) Conference, Madrid, Spain.

Kim, K. H. (2011). The creativity crisis: The decrease in creative thinking scores on the Torrance Tests of Creative Thinking. *Creativity Research Journal, 23*(4), 285–295.

Kramer, M., Tallant, K., Goldberger, A., & Lund, F. (2015). The global STEM paradox. *New York Academy of Sciences*. Accessed at https://nyas.org/media/15805/global_stem_paradox.pdf on May 5, 2018.

Lachapelle, C. P., & Cunningham, C. M. (2014). Engineering in elementary schools. In S. Purzer, J. Strobel, & M. Cardella (Eds.), *Engineering in pre-college settings: Synthesizing research, policy, and practices* (pp. 61–88). Lafayette, IN: Purdue University Press. Accessed at https://eie.org/sites/default/files/research_article/research_file/lachapelle_cunningham_2014_elementary_engineering.pdf on October 29, 2018.

Lai, E., DiCerbo, K., & Foltz, P. (2017). *Skills for today: What we know about teaching and assessing collaboration*. London: Pearson.

Larmer, J. (2015). Gold standard PBL: Public product. *Buck Institue for Education*. Accessed at www.pblworks.org/blog/gold-standard-pbl-public-product on May 28, 2019.

Larmer, J., & Mergendoller, J. R. (2011). Main course not dessert. *Buck Institute for Education*. Accessed at www.bie.org/object/document/main_course_not_dessert on February 4, 2019.

Levi Strauss. (2015). *The life cycle of a jean: Understanding the environmental impact of a pair of Levi's 501 jeans*. Accessed at http://levistrauss.com/wp-content/uploads/2015/03/Full-LCA-Results-Deck-FINAL.pdf on October 17, 2018.

Lucas, B., Hanson, J., & Claxson, G. (2014). *Thinking like an engineer: Implications for the education system, summary report*. London: Royal Academy of Engineering. Accessed at https://raeng.org.uk/publications/reports/thinking-like-an-engineer-implications-summary on May 22, 2018.

MacKay, R. F. (2013). Playing to learn: Panelists at Stanford discussion say using games as an educational tool provides opportunities for deeper learning. *Stanford News*. Accessed at https://news.stanford.edu/2013/03/01/games-education-tool-030113 on October 11, 2018.

Manis, C. (2012). *Quick peer revaluation form*. Accessed at www.lapresenter.com/coopevalpacket.pdf on July 2, 2019.

Mansilla, V. B., & Jackson, A. W. (2014). Educating for global competence: Redefining learning for an interconnected world. In H. H. Jacobs (Ed.), *Mastering global literacy: Contemporary perspectives on literacy* (pp. 5–29). Bloomington, IN: Solution Tree Press.

Markle, W. H. (1966). The manufacturing manager's skills. In R.E. Finley & H.R. Ziobro (Eds.) *The Manufacturing Man and His Job*. New York: American Management Association.

Martin, A. J., & Dowson, M. (2009). Interpersonal relationships, motivation, engagement, and achievement: Yields for theory, current issues, and educational practice. *Review of Educational Research, 79*(1), 327–365.

Maxwell, J. C. (2007). *Failing forward: Turning mistakes into stepping stones for success*. New York: Nelson.

May, C. (2012, March 6). The inspiration paradox: Your best creative time is not when you think. *Scientific American*. Accessed at https://scientificamerican.com/article/your-best-creative-time-not-when-you-think on September 4, 2018.

McFadden, C. (2017). *The origin of the word 'engineering'*. Accessed at https://interestingengineering.com/the-origin-of-the-word-engineering on February 18. 2019.

Merton, R. K. (1936). The unanticipated consequences of purposive social action. *American Sociological Review, 1*(6), pp. 894–904.

Michalko, M. (2006). *Thinkertoys: A handbook of creative-thinking techniques* (2nd ed.). Berkeley, CA: Ten Speed Press.

Miller, R. K. (2015, May). *Why the hard science of engineering is no longer enough to meet the 21st century challenges*. Accessed at www.olin.edu/sites/default/files/rebalancing_engineering_education_may_15.pdf on July 6, 2018.

Mind Tools. (n.d.). *SCAMPER: Improving products and services*. Accessed at https://mindtools.com/pages/article/newCT_02.htm on August 6, 2018.

Moyer, R., & Everett, S. (2010). What makes a better box? *Science Scope, 33*(6), 64–69.

Mullins, A. (February 2009). *Aimee Mullins: My 12 pairs of legs* [Video file]. Accessed at https://ted.com/talks/aimee_mullins_prosthetic_aesthetics on October 18, 2018.

National Academies of Sciences, Engineering, and Medicine. (2018). *English learners in STEM subjects: Transforming classrooms, schools, and lives*. Washington, DC: National Academies Press. Accessed at https://doi.org/10.17226/25182 on February 5, 2019.

National Academies of Sciences, Engineering, and Medicine. (2019). *Minority serving institutions: America's underutilized resource for strengthening the STEM workforce*. Washington, DC: The National Academies Press.

National Academy of Engineering. (2013). *Educating engineers: Preparing 21st century leaders in the context of new modes of learning—Summary of a forum*. Washington, DC: National Academies Press.

National Academy of Engineering & National Research Council. (2002). *Technically speaking: Why all Americans need to know more about technology*. Washington, DC: National Academies Press.

National Academy of Engineering & National Research Council. (2009). *Engineering in K–12 education: Understanding the status and improving the prospects*. Washington, DC: National Academies Press. Accessed at https://doi.org/10.17226/12635 on May 5, 2018.

National Aeronautics and Space Administration. (n.d.). *Bernoulli's principle: Principles of flight*. Accessed at www.nasa.gov/sites/default/files/atoms/files/bernoulli_principle_k-4.pdf on February 18, 2019.

National Aeronautics and Space Administration–Jet Propulsion Laboratory. (n.d.). *Theodore von Kármán*. Assessed at https://jpl.nasa.gov/jplhistory/learnmore/lm-vonkarman.php on July 12, 2018.

National Girls Collaborative Project. (2018). *The state of girls and women in STEM*. Accessed at https://ngcproject.org/sites/default/files/ngcp_the_state_of_girls_and_women_in_stem_2018a.pdf on December 20, 2018.

National Research Council. (1994). *Learning, remembering, believing: Enhancing human performance.* Washington, DC: National Academies Press.

National Research Council. (2006). *Learning to think spatially.* Washington, DC: National Academies Press. Accessed at https://doi.org/10.17226/11019 on February 5, 2019.

National Research Council. (2012). *A framework for K–12 science education: Practices, crosscutting concepts, and core ideas.* Washington, DC: National Academies Press. Accessed at https://doi.org/10.17226/13165 on February 5, 2019.

National Science Board. (2018). *Science and engineering indicators 2018. NSB-2018-1.* Alexandria, VA: National Science Foundation. Accessed at https://nsf.gov/statistics/indicators on January 11, 2019.

National Science Foundation. (2016). *What is engineering?* [Video file]. Accessed at https://nsf.gov/news/mmg/mmg_disp.jsp?med_id=80126 on October 30, 2018.

National Science Teachers Association. (n.d.). *Uncovering student ideas in science.* Accessed at www.nsta.org/publications/press/uncovering.aspx on September 10, 2018.

Newcombe, N. S. (2010). Picture this: Increasing math and science learning by improving spatial learning. *American Educator, 34*(2), 29–35, 43.

Newcombe, N. S. (2013). Seeing relationships: Using spatial thinking to teach science, mathematics, and social studies. *American Educator, 37*(1), 26–31.

NGSS Lead States. (2013). *Next Generation Science Standards: For states, by states.* Washington, DC: The National Academies Press.

Norman, D. (2013). *The design of everyday things* (Rev. ed.). New York: Basic Books.

Novak, J., & Cañas, A. (2008). *The theory underlying concept maps and how to construct and use them.* Accessed at http://cmap.ihmc.us/docs/theory-of-concept-maps on May 5, 2019.

Oppenheimer, J., Davis, M., Carroll, B., Jr. (Writers), & Asher, W. (Director). (1952). Job switching [Television series episode]. In J. Oppenhheimer, J. (Producer) & Arnaz, D. (Executive Producer), *I love Lucy.* Los Angeles: Columbia Broadcasting System. Accessed at https://vimeo.com/69335239 on October 24, 2018.

Osterwalder, A., & Pigneur, Y. (2010). *Business model generation: A handbook for visionaries, game changers, and challengers.* Hoboken, NJ: Wiley.

Papanek, V. (2009). *Design for the real world: Human ecology and social change.* Chicago: Academy Chicago Publishers.

Parker, F., Novak, J., & Bartell, T. (2017). To engage students, give them meaningful choices in the classroom. *Phi Delta Kappan, 99*(2), 37–41. Accessed at www.kappanonline.org/engage-students-give-meaningful-choices-classroom on March 15, 2019.

Partnership for 21st Century Skills. (2008). *21st century skills, education and competitiveness: A resource and policy guide.* Accessed at https://files.eric.ed.gov/fulltext/ED519337.pdf on February 5, 2019.

Pellegrino, J. W., & Hilton, M. L. (Eds.). (2013). *Education for life and work: Developing transferable knowledge and skills in the 21st century.* Washington, DC: National Academies Press.

Peng, X., Peak, D., Prybutok, V., & Xu, C. (2017). The effect of product aesthetics information on website appeal in online shopping. *Nankai Business Review International, 8*(2), 190–209.

Perkins, D. N. (2014). *Future wise: Educating our children for a changing world.* San Francisco: Jossey-Bass.

Pongrácz, E. (2007). *The environmental impacts of packaging.* Accessed at https://researchgate.net/publication/229796182_The_Environmental_Impacts_of_Packaging on September 20, 2018.

Powell, N. (2011). The information blanket: Baby swaddling blanket. *The Grommet.* Accessed at https://thegrommet.com/information-blanket-that-could-save-a-life on May 18, 2018.

Project Learning Tree. (2017). *32 examples of camouflage in nature.* Accessed at https://plt.org/educator-tips/camouflage-nature-examples on October 25, 2018.

Project Zero. (2003). *Eight studio habits of mind.* Accessed at www.pz.harvard.edu/sites/default/files/eight_habits_of_mind%20.pdf on June 3, 2019.

Queen Elizabeth Prize for Engineering. (2017). *Create the Future Report 2017.* Accessed at http://qeprize.org/research/create-future-report-2017 on May 3, 2018.

Index

A
AAAS. *See* American Association for the Advancement of Science
active learning, 62, 82, 105
active listening, 105
Adaptive Design, 175
African American, 75, 78
airplanes, 20, 120–121
American Association for the Advancement of Science, 22, 24, 176
American Society for Engineering Education, 10
analogies, 126
analyses
    of failure, 11–12, 19, 54–55, 120
    of solutions, 54
Anderson, T. R., 105
appropriate technology, 165
Asia Society, 162
Ask Nature website, 145
AsSeenOnTV, 58
assessments. *See also* project assessments
    formative, 14, 25, 70, 78, 80, 86, 88, 116
    group, 89
    individual, 89
    peer, 89–91
    performance, 81–82, 90
    process versus product, 87–88
    prototype failure effects on, 11
    skills, 88–89
    summative, 70, 78, 82
associative thinking, 23
audience, 83

B
background knowledge, 14, 16
background research, 16, 35
backward design, 21, 24, 187
Benchmarks for Scientific Literacy, 176
Benson, T. A., 19
Benyus, J. M., 141
Bernoulli's principle, 64, 97
Bigelow, K. E., 76
biomimicry, 141–142
Biomimicry Institute, 145
blue water (water footprint), 160
board games, 152–159
bottlenecks, 176
Bouffard, S., 105
brainstorming, 41, 43–44, 67, 110, 114–116
brand identification, 38
Buck Institute for Education, 25. *See also* PBLWorks
Building a Better Box project, 46, 135–138
Building a Structure activity, 62

C
Cañas, A. J., 86
Canva, 58
Cardboard Carnival activity, 118–120
case studies, 120
    Challenger, 18, 19, 120
    S.S. Eastland, 19, 120
Cavera, J., 24, 197
challenge(s)
    building of, 60
    components of, 61–62
    constraints in, 38, 60
    content of, 62–64, 67
    core curricular concepts in, 63–64
    defining of, 33–35
    description of, 25–26
    as ill-defined problems, 33–34
    options for, 61
    problem versus, 44
    process of, 66–70
    for projects, 60–61
    quick builds, 64, 71
    skills of, 64–66
challenge statement, 60–61, 103
change
    description of, 4
    layer-based approach to, 8–9
    sustainability of, 9
Chindogu activity, 43, 112–113
civil engineering, 128
classroom culture
    collaboration in, 17
    description of, 9
    as fail forward zone, 12
Cmap, 86
collaboration
    in classroom culture, 9, 12, 17, 60
    communication and, 122–124
    empathy and, 36
    employment skills, 3
    in group projects, 70–78
    as project skill, 65
    self-assessment of, 90
Colorín Colorado!, 77
Committee on Technological Literacy, 3
communication
    about solutions, 52–53
    collaboration and, 122–124
    in engineering design process, 2–3
    as project skill, 65
Community for Advancing Discovery and Research in Education, 125
compliance education model, 26
concept maps, 86–87
concept mastery, 41
Concord Consortium, 42
constraints. *See also* criteria
    in challenges, 38, 60
    cost-based, 17
    creativity and, 60
    design space and, 36
    development of, 37–38, 153–154
    identifying of, 38
    as limitations, 37
    safety-based, 17
    solution and, 45
Constraints and Criteria activity, 103–105
consumer feedback, 50
content
    of challenges, 62–64, 67
    key timing for, 67
continuous improvement, 4
convergent thinking, 31, 41, 191
Cooper Hewitt Design Museum, 117
core curricular concepts, 63–64
core ideas, 28
corrective actions, 176
cost-based constraints, 17
Council of Chief State School Officers, 80
Couros, G., 195
Creating Innovators, 7
creative problem solving, 7–8
creativity, 23, 27, 43–44, 60, 63, 65
criteria
    definition of, 38
    design space and, 36

235

development of, 38–39, 39
ranking of, 39, 45–46
Criteria Ranking activity, 108–109
critical thinking, 1, 23–24, 37, 65, 70
Cummings, M., 21
curricular content, 41
curriculum
  engineering of, 21
  large-scale projects connected to, 8
  optimization of, 190–192
  reflection on, 183–186
  revision of, 187–189
  student identification of, 40
  unit, 12

D

da Vinci, L., 76
Darling-Hammond, L., 80, 82
DDT, 18
design challenge, 33
design expert, 74
Design Modification Request form, 14, 51
Design of Everyday Things, The, 70
design plan, 14
design ranking, 45
design thinking, 36
Designing a Prosthetic Hand (grades 8–12) project, 171–175
direct instruction, 13
disabilities, students with, 76
Disaster-Resilient Housing project, 128–131, 167–170
distractions, 110
divergent thinking, 31, 38, 41, 43, 110

E

Edison, T., 20, 76
EDP. *See* Engineering design process
Educational Research Center of America, 77
Einstein, A., 76
elementary school
  best practices for, 126–127
  reverse engineering, 127
elementary school projects and activities
  Building a Better Box, 46, 135–138
  Engineering Happily Ever After, 131–132
  English language arts-based, 128–135
  Every Graph Tells a Story, 139–141
  Hidden in Plain Sight—Biomimicry, 142–145
  Huff 'n Puff: Disaster-Resilient Housing, 128–131
  Just Right, 132–135
  mathematics-based, 135–141
  No Words: Pictorial Instructions, 146–147
  overall approach to, 127–128
  STEAM-based, 139, 141–147
Ellen MacArthur Foundation, 136
empathy, 36, 65, 105, 174, 191
end users (customers)
  identification of, 35–36, 40
  pain points of, 60
  prototype testing and, 47–48
engagement, 22, 37
Engels, F., 125
Engineer Girl, 77
engineering
  definition of, 2
  habits of mind, 24
  impact of, on student's lives, 9–10
  optimization of solutions, 17
  workforce diversity in, 75
Engineering a Board Game—Visual Arts and Communication project, 152–159
engineering design
  connections in, 14–15
  definition of, 103
  hallmarks of, 10–21
  justifications for, 14–15
  modification request for, 14, 51
  process of. *See* engineering design process
  real-world connections, 22
  requirements for, 27
engineering design challenges. *See also* challenge(s)
  constraints in, 38
  defining of, 33–35
  description of, 25–26
  as ill-defined problems, 33–34
engineering design culture, 7–8
engineering design process (EDP)
  backward design, 21, 24
  communication in, 2–3
  deconstructing of, 27–58
  definition of, 28
  developing solution step of. *See* solutions
  failure in, 11
  as framework, 29
  know your options steps of. *See* know your options
  know your problem step of. *See* know your problem
  overview of, 1–2
  phases of, 29–31
  rubric for, 83–84
  steps of, 28, 32
  subjects, 2–3, 23
engineering engagement experiences, 12
Engineering Happily Ever After project, 131–132
engineering notebook, 54, 70, 78–80
engineering testing, 11
English language arts-based projects, 128–135
English learners, 76–77
ergonomics engineers, 132
ethnic minority students, 78
Every Graph Tells a Story project, 139–141
expert blind spot, 50

F

fail forward, 10, 12, 120
failure
  analysis of, 11–12, 19, 54–55, 120
  as learning tool, 10–11, 96–102
  real-world examples of, 18–19
  testing until, 99
feedback loop diagrams, 19
Fibonacci sequence, 145
Flight of the Table Tennis Ball activity, 100–102
Floyd, Z., 102
Fluckiger, J., 82
fluency, 43
formative assessments, 14, 25, 70, 78, 80, 86, 88, 116
Friedman, R., 110

G

gallery walks, 56–57, 118
Gantt charts, 67–68
Giertz, S., 112
girls and women in engineering, 77, 78
Glogster, 58
Golding, C., 23
gold-standard project-based learning, 25, 53
go/no go mindset, 31
Google Docs, 78
Grandin, T., 76
graphics, 19
gray water (water footprint), 160
Green Eggs and Ham, 60
green water (water footprint), 160
group assessment, 89
group engineering notebook, 54, 70
group grading, 89
group work, 70
groupthink, 44, 109, 115

H

Hammer, B., 15
Harvard Graduate School of Education, 105
Hawking, S., 76
Hendley, V., 10
Herr, H., 76, 171, 175
Hidden in Plain Sight Project—Biomimicry, 142–145
high school
  best practices for, 150–151
  overall approach to, 151
  overview of, 149–150
  STEAM-based projects for, 151–180
high school projects
  Designing a Prosthetic Hand, 171–175
  Engineering a Board Game—Visual Arts and Communication, 152–159
  How Does Your Production Line Rate?—Mathematics, 175–179
  Where Does All the Water Go? Water Footprint Awareness, 159–162
Hodges, L. C., 71
holistic thinking, 19
hooks, 64, 116
How Does Your Production Line Rate?—Mathematics project, 175–179
How It's Made, 42
Huff 'n Puff: Disaster-Resilient Housing project, 128–131
human needs, 4
human-centered design, 36

I

ideas
  brainstorming of, 44
  core, 28
ill-defined problems, 33–34, 61
in-depth research, 43
individual assessment, 89
Initial Design Plan form, 14, 85, 217–218
innovation
  improving and synthesizing existing ideas and products, 20
  team-based approach to, 4
innovation drivers, 37
Innovator's Mindset, The, 195
inputs, 19
inquiry-based learning, 13
Institute of Civil Engineers, 128
interconnectedness, 23
Intermediate Technology Development Group, 165

intrinsic motivation, 37

J
Jackson, A. W., 162
Jones, S., 105
Just Right project, 99, 132–135
justifications, 14–15

K
Kahn, J., 105
kanban board, 67, 69, 78
Katehi, L., 27
Khan Academy, 42
Kim, K. H., 43
kinetic energy, 100
know your options
　activities for, 110–116
　brainstorming possible solutions, 43–44
　overview of, 32
　research to learn more, 41–43
know your problem
　activities for, 103–109
　challenges, defining of, 33–35
　constraints and criteria, 36–41
　end users, 35–36
　overview of, 32–33
Krebs, E., 76

L
large-scale projects, 8
lateral thinking, 19
Latinx, 75, 78
learning
　assessment and, 80
　failure as tool for, 10–11, 96–102
　inquiry-based, 13
　from mistakes, 18, 184
　project-based, 13, 24–25, 53, 63, 80–81
　research and, 41–43
　skills-based, 24, 64, 89, 105
　student engagement in, 192
　trial-and-error approach to, 15–16
learning experience, 21–22
Learning Policy Institute, 80
LEGO Person activity, 122–124
Likert scale, 47, 184
linking phrases, 86
Luciano, J., 12
LucidPress, 58

M
Making Caring Common, 105
Mansilla, V. B., 162
maps, concept, 86–87
Marcinkiewicz, J., 86
marketing manager, 74
marketing pitches, 57, 118
Marmora, A., 66
Martel, M., 192
materials engineer, 74
mathematics-based projects, 135–141
matrix-based weighted ranking process, 108
May, C., 110
media, 58
messy problems, 190
Michalko, M., 114
middle school
　best practices for, 150–151
　overall approach to, 151
　overview of, 149–150
　STEAM-based projects for, 151–180

middle school projects
　Designing a Prosthetic Hand, 171–175
　Engineering a Board Game—Visual Arts and Communication, 152–159
　How Does Your Production Line Rate?—Mathematics, 175–179
　Where Does All the Water Go? Water Footprint Awareness, 159–162
Miller, R. K., 19
mindset, 11
minority students, 78
mistakes
　learning from, 18, 184
　making of, 10–12
modifications, 50–52, 120–122, 191
ModiFly activity, 65, 120–122
Mullins, A., 171

N
NAE. *See* National Academy of Engineering
Name Your Pain activity, 36, 105–108, 112
NASA, 42
National Academies of Sciences, Engineering, and Medicine, 76
National Academy of Engineering, 3, 24, 27, 77
National Action Council for Minorities in Engineering, 78
National Alliance for Partnerships in Equity, 78
National Geographic Kids, 42
National Girls Collaborative, 77
National Research Council
　core ideas, 28
　engineering as defined by, 2
National Science Board, 75
Native American, 74
Neill, M., 82
Newton's Laws of Motion, 23
Next Generation Science Standards, 2, 30, 77, 122, 149
NGSS. *See* Next Generation Science Standards
No Words activity: Pictorial Instructions, 146–147
no-fail zone, 12
Norman, D., 70
notebook, engineering, 54, 70, 78–80
Novak, J. D., 86
NRC. *See* National Research Council
NSB. *See* National Science Board

O
Olin College of Engineering, 19
online resources, 15
open-ended questions, 14
open-endedness, 17
optimization, of curriculum, 190–192
outputs, 19

P
pain points, 35, 60, 105
Papanek, V., 165
Paper Tower of Power activity, 64, 96–98, 129, 168, 201,
PBLWorks, 90
PBS, 77
peer assessment, 89–91
peer pressure, 9
performance assessments, 81–82, 90
Perkins, D. N., 196
pesticides, 18

PhET Simulations, 42
pictorial instructions, 127, 146
pitch, marketing, 57, 118
post-project summary and reflection, 54–55
Powtoons, 58
Practical Action organization, 165
preassessments, 86
presentations, 54, 56
primacy-recency effect, 67
Prism, 10
problem(s)
　challenge versus, 44
　ill-defined, 33–34, 61
　messy, 190
　Ws of, 34–35
problem solving
　creative, 7–8
　engineering and, 23
　engineering design process and, 29
　skills in, 1
problem-based learning, 120
product development, 70
product manager, 74
Proficiency-Based Assessment: Process, Not Product, 88
proficiency-based grading, 88
project(s)
　assessments of. *See* project assessments
　brainstorming during, 115–116
　building of, 60
　challenge for, 60–61
　designing of, 59–91
　engineering notebook documentation of, 78–80
　hands-on building stage of, 116
　learning goals of, 187
　milestones of, 67
　as performance task, 80
project assessments. *See also* assessments
　designing of, 80–87
　learning and, 80
　structuring of, 80–81
project manager, 74
project planning
　description of, 59, 62
project-based learning, 13, 24–25, 53, 63, 80–81
ProjectEngin, 9, 78
prototype(s)
　consumer feedback testing of, 50
　creating of, 46
　deciding on, 45–46
　description of, 44–45
　development of, 46–50
　end user feedback and appeal testing, 47–48
　failure of, 11
　as formative assessment, 116
　functionality testing of, 47, 49
　modifications to optimize, 50–52
　post-project summary and reflection, 54–55
　purpose of, 46
　ranking of, 45
　reasons for, 46
　reliability and safety testing of, 47–48
　size of, 46
　testing of, 31, 46–50, 190
　as visualization, 46
prototyping
　rapid, 116
　value of, 119

prototyping station, 117
PSA. *See* public service announcement
public service announcement, 117

Q

quick builds, 12, 64, 71, 116

R

rapid prototyping, 116, 118, 134, 137
Ready, Set, Design activity, 117
real-world
    connections, 22, 162–179
    failures, 18
    real-world global challenges, 162–171
    scenarios, 60
    thinking, 191
reflection, 183–186
reinforcement, 88
Remezcla, 78
reports, 54
research
    in-depth, 43
    learning through, 41–43
    time for, 16
    websites for, 42
research project, 42
reverse design space form, 104
reverse engineering, 127
revision, 187–189
role-playing, 53
Roosevelt, F. D., 195
Roschewski, P., 82
Rosling, H., 139
rote memorization, avoiding 9
Royal Academy of Engineering, 10
rubrics, 82–85, 91
single-point rubric, 82, 85, 90
Ruiz-Primo, M. A., 86

S

safety-based constraints, 17
SCAMPER This! activity, 43, 114–115, 201
Schrock, K., 82
Schumacher, E. F., 165
Science Advances, 136
science experiments, 11
Scrum method, 67
Scutieri, S., 63, 110, 180
SDGs. *See* sustainable development goals
secretary effect, 75
Seuss, Dr., 43, 60
Shark Tank, 57
single-point rubric, 82, 85, 90
Sketchy Ideas activity, 117–118
skills
    assessment based on, 88–89
    challenge, 64–66
    in problem solving, 1
skills-based learning, 24, 64, 89, 105
Small Is Beautiful, 165
Smith, A., 165
Smithsonian Institute, 117
Smithsonian National Air and Space Museum, 21
solutions
    analyses of, 54
    brainstorming of, 43–44
    communication about, 52–53
    development of, 32, 44–52, 116–124
    gallery walks of, 56–57
    inputs and, 19
    marketing pitches, 57
    media, 58
    modify to optimize, 50–52
    optimization of, 17, 50–52
    outputs and, 19
    presentations of, 54, 56
    problems created by, 18–19
    prototypes. *See* Prototype(s)
    reports of, 54
Sorby, S. A., 77
standards-based grading, 88
Stars and Stripes (and Dots) activity, 43, 111–112
STEAM-based projects
    for elementary school, 139, 141–147
    for high school, 151–180
    for middle school, 151–180
Steinkuehler, C., 152
STEM
    connections in, 2
    for English learners, 76
    women in, 77
STEMConnector, 3
stock-flow maps, 19
student(s)
    background knowledge, 14
    big-picture look with, 39–41
    direct instruction to, 13
    with disabilities, 76
    engagement of, in learning, 192
    engineering impact on, 9–10
    ethnic minority, 78
    as experts, 43
    reinforcement for, 88
    self-reflection by, 89
student choice, 25–26
student differentiation, 25–26
subassemblies, 176
summative assessment, 70, 78, 82
summative test, 12–13
sustainable development goals, 163–165
Suster, M., 106
Sutton, A., 145
synthesis, 111
systems thinking, 12, 18–19, 66

T

TeachEngineering, 108
Teacher-Written Eduware website, 90
team/teamwork
    ability levels in, 72
    assembling of, 71–73
    English learners on, 76–77
    girls on, 77
    group sizes in, 71
    innovation and, 4
    jobs in, 73–75
    learning experiences of, 72
    mixed groups, 72
    multitasking approach to, 73
    single-sex groups, 72
    student involvement in selection of groups, 72–73
    students with disabilities on, 76
    synergy increased with, 70–78
    underrepresented students on, 75–78
technological literacy, 3
TED Talks, 10, 39, 64, 76, 112, 139, 171
TED-Ed, 42
testing until failure, 99
Thinkertoys, 114
thinking
    convergent, 31, 41, 191
    divergent, 31, 38, 41, 43, 110
    holistic, 19
    lateral, 19
    real-world, 191
    systems, 12, 18–19, 66
    transdisciplinary, 22–24
Three-Legged Stool activity, 98–100
to-do lists, 78
Tomaswick, L., 86
Tomita, M., 86
transdisciplinary thinking, 22–24
Treuer, D., 124, 196
trial and error
    learning through, 15–16
    modification and, 52
    projects based on, 44
    prototype results and, 47
Truesdell, A., 175

U

unintended consequences, 18
United Nations, 163–165
universal design principles, 76
University of Otago, 23
un-useless invention, 112

V

Vanides, J., 86
vision
    direction and, 21–26
    real-world connections and, 22
visual aids, 19
von Kármán, T., 27
voting slips, 108

W

Wagner, T., 7
water footprint, 159–160
Waters Foundation, 19
website resources, 15
Weissbourd, R., 105
What Is Engineering? video, 10
Where Does All the Water Go? Water Footprint Awareness project, 159–162
Wood, G. H., 82
workforce diversity, 75
World Economic Forum, 136
World Health Organization, 18
"World Needs All Kinds of Minds, The," video 76
Wright, O., 20
Wright, W., 20
Ws, five 34, 103

Y

yield, 176
Yin, Y., 86

Z

Zaccone, R., 193

Growing Tomorrow's Citizens in Today's Classrooms
Cassandra Erkens, Tom Schimmer, and Nicole Dimich Vagle

For students to succeed in today's ever-changing world, they must acquire unique knowledge and skills. Practical and research-based, this resource will help educators design assessment and instruction to ensure students master critical competencies, including collaboration, critical thinking, creative thinking, communication, digital citizenship, and more.
BKF765

Ready for Anything
Suzette Lovely

Effective teaching and learning must reflect what's happening technologically, socially, economically, and globally. In *Ready for Anything*, author Suzette Lovely introduces four touchstones that will invigorate students' curiosity and aspirations and prepare them for college, careers, and life in the 21st century.
BKF848

Fifty Strategies to Boost Cognitive Engagement
Rebecca Stobaugh

Transform your classroom from one of passive knowledge consumption to one of active engagement. In this well-researched book, Rebecca Stobaugh shares fifty strategies for building a thinking culture that emphasizes essential 21st century skills—from critical thinking and problem-solving to teamwork and creativity.
BKF894

Future-Focused Learning
Lee Watanabe-Crockett

When educators embrace student-centered learning, classrooms transform, learning comes alive, and outcomes improve. With *Future-Focused Learning*, you will discover ten core shifts of practice—along with simple microshifts—that will help take the great work you are already doing and make it exceptional.
BKF807

Different Schools for a Different World
Scott McLeod and Dean Shareski

Explore six key arguments for why educators must approach schooling differently: (1) information literacy, (2) the economy, (3) learning, (4) boredom, (5) innovation, and (6) equity. Learn how schools are tackling each argument head-on to prepare students for the demands of the global world.
BKF729

Solution Tree | Press  Visit SolutionTree.com or call 800.733.6786 to order.

Wait! Your professional development journey doesn't have to end with the last pages of this book.

We realize improving student learning doesn't happen overnight. And your school or district shouldn't be left to puzzle out all the details of this process alone.

No matter where you are on the journey, we're committed to helping you get to the next stage.

Take advantage of everything from **custom workshops** to **keynote presentations** and **interactive web and video conferencing**. We can even help you develop an action plan tailored to fit your specific needs.

Let's get the conversation started.

Call 888.763.9045 today.

SolutionTree.com